Stage 1 Design

Electrical Installation Series – Intermediate Course

M. Doughton
E. G. Stocks

Edited by Chris Cox

THOMSON

Australia · Canada · Mexico · Singapore · Spain · United Kingdom · United States

THOMSON

Stage 1 Design

British Library Cataloguing-in-Publication Data
A catalogue record for this book is available from the British Library

ISBN-13: 978-1-86152-715-8
ISBN-10: 1-86152-715-2

First published 2001 by Thomson Learning
Reprinted 2002, 2003, 2004 and 2005 by Thomson Learning

Printed in Croatia by Zrinski d.d.

About this book

"Stage 1 Design" is one of a series of books published by Thomson Learning related to Electrical Installation Work. The series may be used to form part of a recognised course or individual books can be used to update knowledge within particular subject areas. A complete list of titles in the series is given below.

Electrical Installation Series

Foundation Course

Starting Work
Procedures
Basic Science and Electronics

Supplementary title:
Practical Requirements and Exercises

Intermediate Course

The Importance of Quality
Stage 1 Design
Intermediate Science and Theory

Supplementary title:
Practical Tasks

Advanced Course

Advanced Science
Stage 2 Design
Electrical Machines
Lighting Systems
Supplying Installations

Acknowledgements

The authors and publishers gratefully acknowledge the following illustration sources:

Bussmann Division, Cooper (U.K.) Ltd. (pp. 23 and 64); Hager Powertech (p. 6).

Every effort has been made to trace all copyright holders but if any have been inadvertently overlooked, the publishers will be pleased to make the necessary arrangements at the first opportunity.

Study guide

This studybook has been written to enable you to study either in a classroom or in an open or distance learning situation. To ensure that you gain the maximum benefit from the material you will find prompts all the way through that are designed to keep you involved with the subject. The book has been divided into parts each of which may be suitable as one lesson in the classroom situation. Certain parts of this book may be combined in one lesson period but this will depend upon the duration of the lesson. However if you are studying by yourself the following points may help you.

☞ Work out when, and for how long, you can study each week. Complete the table below and from this produce a programme so that you will know approximately when you should complete each chapter, the project and end tests. Your tutor may be able to help you with this. It may be necessary to reassess this timetable from time to time according to your situation.

☞ Try not to take on too much studying at a time. Limit yourself to between 1 hour and 2 hours and finish with a task or the self assessment questions (SAQ). When you resume your study go over this same piece of work before you start a new topic.

☞ You will find the answers to the questions at the back of the book but before you look at the answers check that you have read and understood the question and written the answer you intended.

☞ A project is included in this book. You will find details in the Appendix on p.131. A worked example (Circuit 1) can be found in the text at the end of Chapters 3, 4, 5, and 6. and you should then complete the appropriate stage for Circuit 2. Circuits 3 and 4 should be completed after Chapter 6. At the end of Chapters 7 and 8 are questions which refer to the specification and plans of the small factory in the project. An "end test" covering the material in Chapters 7 and 8 is included so that you can assess your progress.

☞ Tasks are included where you are given the opportunity to ask colleagues at work or your tutor at college questions about practical aspects of the subject. These are all important and will aid your understanding of the subject.

☞ It will be helpful to have available for reference a current copy of BS 7671 and IEE Guidance Notes 1 "Selection and Erection" and IEE Guidance Notes 3 "Inspection and Testing" when studying this book.

☞ Your safety is of paramount importance. You are expected to adhere at all times to current regulations, recommendations and guidelines for health and safety.

Study times					
	a.m. from	to	p.m. from	to	Total
Monday					
Tuesday					
Wednesday					
Thursday					
Friday					
Saturday					
Sunday					

Programme	Date to be achieved by
Chapter 1	
Chapter 2	
Chapter 3	
Chapter 4	
Chapter 5	
Chapter 6	
Chapter 7	
Chapter 8	
Chapter 9	
Project (Circuits 3 and 4)	
End test	

Contents

1

Supply and Protection

Before you start work on this subject, complete the exercise below to ensure that you remember what you have already learned.

Which organisation is responsible for local electricity transmission?

What is the nominal supply voltage to a residential house likely to be?

Where does the consumer's installation begin?

On completion of this chapter you should be able to:

- ◆ state the standard generation, transmission and distribution voltages in use in the UK
- ◆ explain why it is important to have a network grid system
- ◆ give reasons for the different voltages used to supply the various types of industrial, commercial and domestic installations
- ◆ explain the need for balancing the loads evenly across three phases
- ◆ describe the consumer's distribution system relative to:
  – isolation and switching,
  – overcurrent protection and
  – earth fault protection
- ◆ give examples of where protection is incorporated at the main intake position
- ◆ describe how to safely isolate a section of an installation to recommended procedures
- ◆ calculate the assumed load when applying diversity

Part 1

Electricity has become part of our everyday lives. We accept it as always being there and expect it to perform all the functions we require of it.

It is generated in power stations and transmitted over hundreds of miles to where it is needed. The pylons used to carry the transmission cables have almost become part of the landscape. However, if we look carefully and start comparing them we find that they vary in height and style and this is no accident for they carry different voltages and different loads are involved.

The loads on the cables can be critical, especially if they become out of balance.

Before the electricity gets into the factory or home, it has to go through a distribution network that is both safe and practical.

Some of the material in this chapter has been covered at the foundation course level but it is important to use this as a basis for the more advanced work in this intermediate level.

The supply

Whenever we go to switch on electrical equipment, we expect the supply to be there. To do this, electrical generators must be working 24 hours a day, every day of the year because mains electricity is not stored and it is generated all the time. To make sure there is enough power available for any anticipated load, generators across the country have to be running regardless of whether the supply is used.

Not only do the generators have to be running but there must be an electrical connection between the generator and where the electrical energy is to be used. The cables making this connection often have to be hundreds of miles long. In order to make the most efficient use of these cables the voltage is often varied several times before the supply reaches the consumer.

Generation to transmission

The output voltage of a generator set in a power station will not exceed 25 000 volts a.c., and with older generators it may be considerably less.

For transmission purposes the generated output is stepped up to a maximum of 400 000 volts (400 kV) or 275 000 volts (275 kV) in some situations.

The grid system

The system of high voltage transmission is known as the Grid System and consists of a lattice work of cables which connect the power stations and large load areas together.

Power stations are not generally situated where the highest loads are. This means that there must be cables from the power stations to all of the load points. There are also interconnections between power stations so that they can cover for each other.

Nationally not all peak loads appear at the same time and a power station in one area may be called on to supply power to another. From time to time generators or power stations have to be closed down for maintenance and repair and when this happens other power stations must cover the total demand.

Over the last 30 years the trend has been to build bigger and more efficient power stations and close the smaller, less economic ones down.

In addition to the UK grid system there are cross channel links with the French supply system. These are underwater cables going from the Kent coast to North East France. Cultural differences and a different working practice make energy exchange possible as the peak loads occur at different times. The UK is therefore able to import power from France when the demand requires it. It also allows the opposite to take place when the French demand is high. For practical reasons these cross channel cables are supplied with d.c. and this is converted to a.c. at each end.

High voltage transmission

Cables transmitting power over long distances need to be kept as small as possible for practical and economic reasons. Conductors with large cross-sectional areas would involve far bigger pylons to take the extra weight and would probably require more of them. By using high voltages, conductors are kept as small as reasonably possible.

Let's look at an example of how this works.
If we assume the average household has a load of about 15 kW then a town with 10 000 houses would have a demand of

$$10\,000 \times 15 = 150\,000 \text{ kW.}$$

If this demand is then split over three phases, each phase would supply approximately 50 000 kW.

If this was supplied by cables at 230 V the conductor would need to be capable of carrying

$$\frac{50\,000 \text{ kW}}{230 \text{ V}} = \frac{50\,000\,000 \text{ W}}{230 \text{ V}} = 217\,391 \text{ A}$$

So a very large cable would be required.

However if this same load is supplied by cables at 400 kV the conductors will only need to be capable of carrying

$$\frac{50\,000 \text{ kW}}{400 \text{ kV}} = \frac{50\,000\,000 \text{ W}}{400\,000 \text{ V}} = 125 \text{ A}$$

A single 400 mm^2 overhead transmission cable which could be used on 400 kV supplies has a current rating of 650 A. These cables are usually bunched in twos or fours on the pylons for more efficient transmission of energy.

Loads on a.c. supplies

To simplify the load calculation, watts and kilowatts have been used, however, in practice on a.c. supplies the loads would be measured in kVA and not kW. When using d.c. supplies power, in watts, can be calculated from the volts multiplied by the amperes. When alternating currents are used it is only on purely resistive loads that this is true. All equipment that uses coils of wire associated with an iron core has what is known as reactance and this results in the voltage and current being out of phase with each other.

If a motor circuit was connected with a voltmeter, ammeter and wattmeter the readings on the voltmeter multiplied by those on the ammeter would not equal that of the wattmeter. In fact they would be greater. The factor that makes up the difference can be calculated from watts divided by VoltAmperes and this is known as the power factor.

The result of this is that a.c. "power" may apparently be measured in two quantities:
- kW when load is purely resistive
- kVA when the load is not purely resistive

In fact there is a third measurement known as kVAr which relates to the reactive part of a load. These three factors can be represented by a scaled triangle (Figure 1.1).

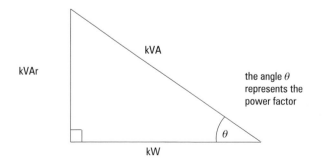

Figure 1.1   kW is the base
kVAr is usually vertical
kVA is the line drawn between the two and is always the highest value.

kVA can be determined by drawing the triangle to scale or by using the formula

$$kVA^2 = kW^2 + kVAr^2$$

$$or \quad kVA = \sqrt{kW^2 + kVAr^2}$$

Transmission and distribution

Although the modern distribution system can sometimes appear to be very complex, it is far better than was available in the past.

For example in London in 1919 there were eighty separate supply undertakings with seventy different generating stations, fifty different supply systems operating at twenty-four different voltages and ten different frequencies. Since those times voltages and frequencies have been standardised.

The supply of electrical energy at voltages above 33 kV, which forms part of the National Grid, is referred to as the Transmission Network. We shall consider the distribution system which operates at 11 kV and 400/230 V. These are known as secondary and tertiary distribution. The distribution system above 11 kV and below 33 kV is known as the primary distribution.

The Electricity Suppliers have a legal responsibility to keep the supply within certain limits. Following voltage standardisation in Europe these are:
- for voltage a nominal supply of 400/230 V, + 10%, –6%;
- frequency must not be more or less than 1% of 50 Hz over a 24-hour period.

The public distribution system can be split into three main sections
- industrial
- commercial and domestic
- rural

The reason for separating them in this way is because of the differences in power consumption and the remoteness of rural supplies. Distribution cables are laid underground wherever possible. With the exception of rural areas, almost all 11 kV and 400/230 V cables are buried underground and a large number of the higher voltage cables are now buried.

Industrial consumers may take their supply at 33 kV and in some cases 132 kV. If an industrial estate consists of smaller units, the estate will have a substation supplied with 132 kV or 33 kV. The transformer will then step-down the voltage to 11 kV for further distribution. In the case of very small units the substation may transform directly down to 400/230 V.

Large commercial premises will have their own substation transformer fed at 11 kV, which will step down the voltage to 400/230 V for internal distribution, whilst smaller commercial and domestic consumers are usually supplied at 400/230 V.

The 11 kV input to the transformer will be connected in delta whereas the 400/230 V output will be a star arrangement (Figure 1.3). To supply the delta connected windings a three-phase three-wire system is used, with no neutral conductor. The star connected output uses a three-phase four wire connection with the centre point of the star being the neutral which is connected to earth.

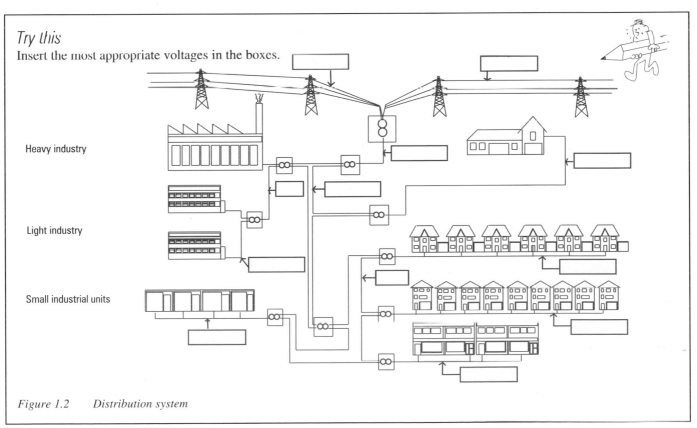

Try this

Insert the most appropriate voltages in the boxes.

Heavy industry

Light industry

Small industrial units

Figure 1.2   Distribution system

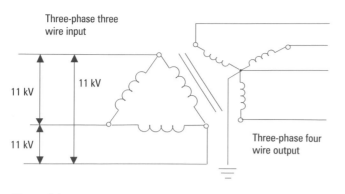

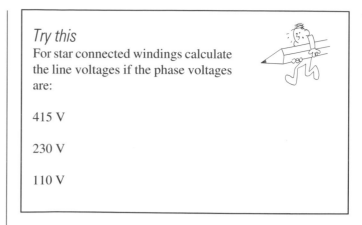

Figure 1.3

In delta the voltage across each of the lines (line voltage) is the same as the transformer winding, whereas a transformer winding connected in star will give us a line voltage between phases of 400 V (U) and 230 V (U_0), phase voltage, between any phase and the neutral star point (Figure 1.4).

Note: The colour identification of cables in the UK is Red, Yellow and Blue.

Try this

For star connected windings calculate the line voltages if the phase voltages are:

415 V

230 V

110 V

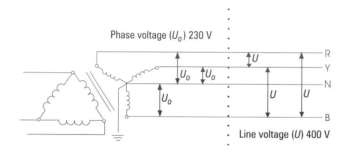

Figure 1.4   *Voltages available from a star connected transformer winding*

The relationship between the phase voltage and the line voltage for a star connected winding is:

$$\text{phase voltage} = \frac{\text{line voltage}}{\sqrt{3}}$$

$$\text{line voltage} = \text{phase voltage} \times \sqrt{3}$$

$$\text{where} \quad \sqrt{3} = 1.73$$

Example:

The phase voltage is 230 V for most domestic premises. The line voltage for this supply is

$$230 \times \sqrt{3} = 230 \times 1.73$$
$$= 400 \text{ V}$$

Similarly a 400 V line voltage gives a phase voltage of:

$$\frac{400}{\sqrt{3}} = \frac{400}{1.73} = 230 \text{ V}$$

Load currents in 3 phase circuits

It is important to recognise the relationships of the currents in star and delta connected windings. In star connected the current through the line conductors is equal to that flowing through the phase windings, as shown in Figure 1.5.

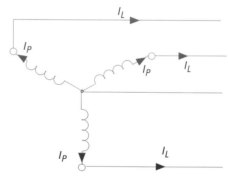

Figure 1.5

However in the delta connected winding this appears to be more complex. The line current, when reaching the transformer winding, is split into two directions so that two phase windings are each taking some current. As each of the phases is 120° out of phase with the others, and the current is alternating, each line conductor acts as a flow and return.

The arrows shown on Figure 1.6 only give an indication as to the current distribution from each line and **all of these currents would not be flowing in the directions shown at the same time.**

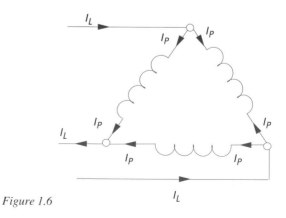

Figure 1.6

The currents through the phase windings are:

$$I_P = \frac{I_L}{\sqrt{3}}$$

or $\quad I_L = I_P \times \sqrt{3}$

Example:
If the line current is 100 A the phase current is

$$I_P = \frac{100}{\sqrt{3}} = \frac{100}{1.73}$$

$$= 57.8 \text{ A}$$

Three phase balanced loads

All transmission and primary distribution is carried out using a three-phase system. It is important that each of the phases carries about the same amount of current.

Three-phase motors have equal windings and each phase is the same. Therefore the conductors carry the same current, and these automatically create a balanced load situation.

For domestic areas the output of the star connected transformer is 400/230 V. All premises are generally supplied with a phase and neutral at 230 V. To try to balance the loads on each of the phases, houses may be connected as shown in Figure 1.7. If it was possible to load all of the phases exactly the same the current in the neutral would be zero.

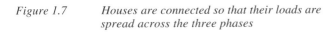

Figure 1.7   Houses are connected so that their loads are spread across the three phases

High voltage transmission
For transmission purposes the generated output, which will not exceed _____ volts a.c., is stepped up to a maximum of _____ kV.

The system of high voltage transmission is known as the _____ _____.

A single 400 mm² overhead transmission cable which could be used on 400 kV supplies has a current rating of _____ A.
For more efficient transmission of power these cables are usually bunched in _____ or _____.

Distribution
The supply of electrical energy at voltages above 33 kV is referred to as the _____ _____.

Primary distribution is the system which operates at _____ kV.

Secondary distribution is the system which operates at _____ kV.

Tertiary distribution is the system which operates at _____ kV.

The 11 kV input to the transformer will be connected in _____, whereas the 400/230 V output will be a _____ arrangement.

The relationship between the phase voltage and the line voltage for a star connected winding is given by the formula:

Load currents in 3 phase circuits
In star connected windings the current through the _____ _____ is equal to that flowing through the _____ _____.

In the delta connected winding the line current, when reaching the transformer winding, is split into ____ _____ .
As each of the phases are _____ out of phase with each other and the current is _____ each line conductor acts as a _____ and _____.

Three phase balanced loads
For domestic areas the output of the star connected transformer is ____ / ____ V. All premises are supplied with a _____ and _____ at 230 V. If it was possible to load all of the phases exactly the same the current in the neutral would be _____ A.

Part 2

As we have seen, the distribution of electrical energy is tailored to the needs and location of the consumer in relation to the supply source. The use of transformers to determine the distribution system voltage allows the supplier to size cables and control gear to give maximum network efficiency and convenience to the customer. Once the supply enters the consumers' premises connection is made to the customer's own distribution system and final circuits. These will vary depending on the type of consumer, the energy requirement and the type of equipment being used, so we shall consider typical final circuit arrangements for the most common groups of consumers.

Domestic

Domestic consumers generally receive their supply at 230 V, single phase, from the supply company, although in areas where electric space heating is employed a second or third phase may be supplied. This is because, depending on the size of the heating load, it may be necessary to distribute the current demand across the phases.

The final distribution circuits for the domestic installation, lighting and power circuits, operate at 230 V, single phase throughout. Some items of equipment, for specific locations or in order to provide specialist control or effects, are supplied at a lower voltage. These items of equipment will generally operate at extra low voltage, below 50 V, and are supplied via their own local transformer.

Each final circuit must be controlled by a protective device rated at a suitable level to supply the load of the circuit which it protects. The requirements for the protection of the final circuits are covered later in this chapter. Each final circuit must have its own protective device and the "distribution board" for modern domestic installations is generally a consumer unit containing miniature circuit breakers or fuses. Other protective devices, such as residual current circuit breakers may also be incorporated and again these will be considered at a later stage. A typical domestic consumer unit is shown in Figure 1.8, and the rating of the devices will vary between say a 6 A device for the protection of the lighting circuits through to 45 A devices for the larger power using equipment such as large electric showers, cookers and the like. A typical domestic schematic diagram is shown in Figure 1.9.

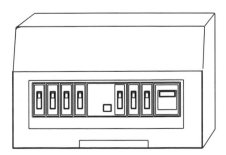

Figure 1.8   Typical domestic consumer unit

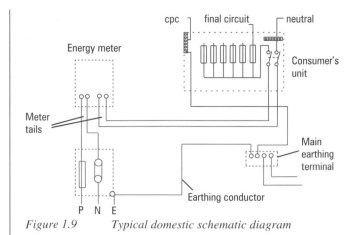

Figure 1.9   Typical domestic schematic diagram

Commercial

The arrangement of final circuits in small commercial premises may be very similar to that of the domestic installation. However, larger commercial buildings will have a considerably bigger floor area and a greater power requirement. The same consideration needs to be given to the economic use of conductor sizes and effective utilisation of equipment within the consumer's installation as is given to the supply network. As a result it is common to install "distribution circuits" within the installation. In a multi-storey building, for example, a distribution circuit may be installed to each floor. These would generally comprise a means of isolation and protection at the intake position, a large cross-sectional area cable to a convenient point on the appropriate floor and a distribution board containing the final circuit protection devices for the circuits on that level. It is common to try to locate these distribution boards as close to the centre of the area as possible in order to minimise the length of the final circuit cables. It is common for the consumer's distribution circuit to be at 400 V, three phase and neutral with the lighting and power circuits for the floor area being supplied at 230 V single phase. Again special areas, or the requirement for sophisticated controls, may require the use of extra low voltage systems being installed.

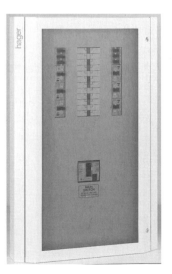

Figure 1.10   Typical commercial installation

Reproduced with kind permission from Hager Powertech Ltd

A typical schematic layout for a commercial installation is shown in Figure 1.11. Remember that a similar approach may be taken to a single storey building which covers a large floor area with the distribution boards being placed at convenient locations around the building.

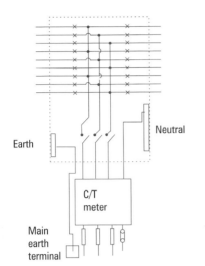

Figure 1.11   Schematic layout for a commercial installation

Industrial

Small industrial units will often receive their supply at 400/230 V from the supply company and their final circuit arrangements may be very similar to those of the commercial installation above. Many industrial applications involve the use of equipment with a high power requirement, which is often achieved most economically with three phase equipment. It is quite common to find that the general lighting and power in such installations are at 230 V single phase and that 400 V three phase final circuits (Figure 1.12) are installed to supply particular items of equipment such as motors.

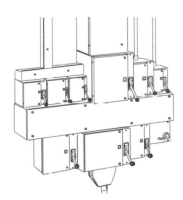

Figure 1.12   Three phase industrial system

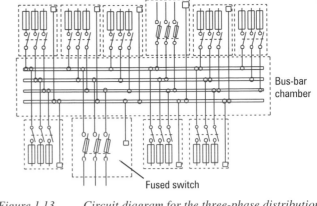

Figure 1.13   Circuit diagram for the three-phase distribution system in Figure 1.12.

Large industrial consumers may take their supply at high voltage, with very large consumers having supplies at 132 kV and above. Where the supply is at 11 kV or above the consumer will need to provide their own sub station transformer and distribution equipment. There will also be a need for the consumer to engage suitably qualified and competent engineers to maintain their system. Loss of even a part of the system could seriously disrupt the consumer's activities. In essence the consumer's distribution network can, for our purposes, be regarded as being the same as that of the supply company. The consumer's network is used to economically provide an electrical supply to the parts of the installation where it is required. In some instances the consumer may have equipment which operates at voltages in excess of 400 V three phase, but we shall not consider those here.

Having been distributed through the consumer's own network and transformer(s) the supply for each area is derived from an intake position within the consumer's premises. Once again this may be further distributed as in the commercial and smaller industrial installations described above.

Agricultural

Agricultural installations are generally located in remote areas, surrounded by agricultural land. As a result they are some distance from the distribution network of the supply company. It is common for such installations to be supplied via an overhead line network as the cost of burying supply cables over the distances involved is uneconomic. Because of the remote location it is often economical for the supply company to install a pole-mounted transformer local to the installation,

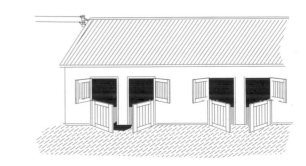

Figure 1.14

thus reducing the losses on the supply network. Farm installations often receive their supply at 400/230 V and this is then distributed to the farm buildings (Figure 1.14), often by the use of overhead conductors. The general lighting and power will be at 230 V single phase but, as with the smaller industrial equipment, many of the items of machinery will require 400 V three phase supplies.

Each of the above installations will have its own unique requirements and it is not possible to consider them all here. We can however generalise on the requirements for the different types of final circuit and consider typical applications. We can categorise the basic final circuits into some specific areas and consider the requirements for each.

Heating

Final circuits in this category will include those supplying space heating, water heating and cooking applications. As a general rule these final circuits will be supplying equipment which will draw considerable current, and require large cross-sectional area cables to supply it. By supplying such circuits at voltages greater than 230 V single phase, 50 Hz, the current required to produce the same power output can be considerably reduced, as we found earlier in this section. In domestic premises it is not often practical or cost effective to supply this equipment at above 230 V and as a result these circuits generally require the largest cables. Installations which receive their supply at 400 V and above will benefit from supplying their large heating loads at 400 V three phase and above.

Power

General power outlets are provided at 230 V and these are common to most installations. The type of outlet may vary dependent upon the type of equipment and the intended use of the circuit, but the 13 A socket outlet appears in almost every installation. Larger equipment which requires more power will be supplied by a dedicated final circuit. Some industrial applications require the installation of 400 V, three phase or three phase and neutral socket outlets. These are often used to supply portable or transportable equipment which would otherwise require a much larger single phase circuit.

Lighting

Lighting is generally supplied at 230 V single phase, and the type of lamp used determines the light output, with specialist lamps being employed for applications such as street lighting (Figure 1.15), car parks, sports halls and the like. Some applications require the lighting to be supplied at extra low voltage and some specialist locations may require this to be at safety extra low voltage. Both these being 50 V a.c. or less, the SELV being supplied through a safety isolating transformer to BS 3535.

Control circuits

Many electrical installations now include equipment which requires sophisticated control circuitry. The domestic dimmer switch, operating at 230 V, 50 Hz is perhaps one of the most basic controls. Industrial and commercial installations use equipment

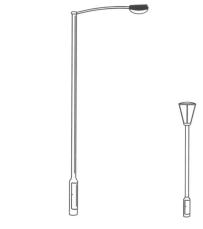

Figure 1.15

which requires an elaborate control system. The commercial installation may require controls on air conditioning and heating systems and there may be a need for environmental control within an installation. Many companies now use energy management systems which monitor and control all aspects of energy consumption. The systems, whilst seemingly expensive can, in the longer term offer considerable cost savings for the company. The industrial consumer will often require complex control over the production process resulting in considerable savings in both manpower and reduced wastage.

These systems often involve the use of microprocessors and logic controls and whilst the equipment being controlled may have a large power consumption the actual control processes require minute power levels. The current requirements are very low and as a result the use of extra low voltage equipment in electronics is most suitable. The insulation values with voltages as low as 12 V a.c. or d.c. and the small current requirement means that small cables can be used for the interconnection of the control devices. Control circuits are generally supplied via panel-mounted transformers, for the a.c. devices, or transformer rectifiers for the d.c. equipment. The control circuits are often supplied through special devices to prevent fluctuations in the supply or interference from the supply system from interfering with the control system.

Alarm systems

In much the same way as the control circuits, alarm systems operate on very small power requirements, as they basically rely on a change of state in an electronic circuit to cause the operation of an alarm. The majority of systems now use electronic equipment throughout and, like our control systems, operate at voltages as low as 12 V. Many systems are available, some operating on a.c. and some on d.c. Generally the supply to the system is derived from a 230 V a.c. supply and the equipment transforms and rectifies this to provide the extra low voltage for the alarm system. Many alarms operate through internal supplies which incorporate a device allowing the system to be supplied from a separate, usually self contained, d.c. source. This source, normally a battery, is maintained in a fully charged state by the main supply. However in the event of a main supply failure the battery is able to take over and run the system.

Distribution circuits

The final distribution circuit for the domestic installation, lighting and _____ circuits, operates at _____V, _____ phase throughout.

Extra low voltages to supply some specialist equipment are generally below _____ V and are supplied via their own local _____.

The distribution board for modern domestic installations is generally a _____ _____ containing _____ _____ _____ or _____.

The distribution circuits in a large commercial arrangement occupying a multi-storey building may be installed on _____ _____. These would generally comprise a means of _____ and _____ at the intake position, a large cross-sectional cable to a convenient point on the appropriate floor and a distribution board containing the _____ _____ _____ _____ for the circuits on that level.

Large industrial consumers which take their supply at 11 kV or above will need to provide their own ____ _____ _____ and _____ equipment.

Farm installations often receive their supply at ____ / ____ V and this is often distributed to farm buildings by the use of _____ _____.

The majority of alarm systems now use electronic equipment and operate at voltages as low as ____ V.

Many alarms operate through internal supplies which incorporate a device allowing the system to be supplied from a separate _____ source. This source is normally a _____ which is maintained in a fully _____ state by the main supply.

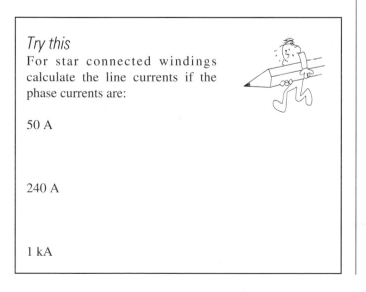

Try this

For star connected windings calculate the line currents if the phase currents are:

50 A

240 A

1 kA

Part 3

The three main requirements for protection

Within every part of an installation there are at least three main requirements that must be considered.

These are the need for:
- isolation
- automatic protection against overload
- automatic disconnection in the event of current leakage to earth

These are all concerned with safety and need to be installed for protection not just from electric shock, but also from fire, burns or injury from mechanical equipment which is electrically activated.

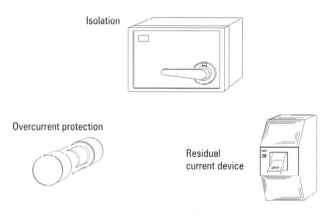

Figure 1.16   Typical devices for protection

Isolation

The term isolation, in this context, means the cutting off of the installation, or circuit, from all sources of electrical supply. This is necessary to allow maintenance or repair work to be carried out without fear of shock from direct contact, causing a short circuit between live conductors or between live conductors and earth. Short circuit and earth faults will be dealt with in more depth later in this workbook. At this point it may be a good idea to look at what is meant by "direct contact". If a person comes into contact with parts of an installation that are intended to be live under normal conditions then they are said to have received an electric shock through direct contact.

During normal use live parts are covered with insulation or enclosed in cabinets or boxes so that it is not possible to make direct contact. If we are to, for example, change a socket outlet we must expose the bare conductors and terminals during this operation, so running the risk of electric shock by direct contact.

One of the main reasons that isolators are installed is in order that they can be switched off before equipment is maintained or repaired so that there are no exposed live parts (Figure 1.17).

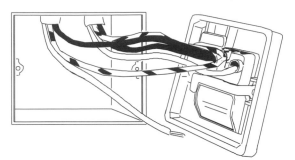

Figure 1.17   Circuits should be isolated before any alterations are made

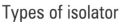

Remember
An isolator must cut off an electrical installation, or part of it, from every source of electrical energy.

Types of isolator

There are a number of types of isolator each performing similar functions but with different numbers of contacts. The number of conductors broken gives the isolator its name. These names tend to be rather long especially to write out each time and so are often abbreviated to letters. This gives us the following:

SP   – single-pole

DP   – double-pole

TP   – triple-pole

TP&N   – triple-pole and neutral

Try this
Mark under each item in the list below the type of isolator required.

1.   A single phase lighting circuit

2.   A three-phase motor

3.   A three-phase and neutral distribution board

4.   A single-phase water heater

Isolators must also break all of the contacts at the same time. To do this the switches are linked together by insulating material so ensuring that all conductors are made or broken simultaneously without shorting each other out. The neutral in TP & N is not switched but it is usually connected via a bolted link connection.

Location of isolators

Now that we have considered the types of isolators we need to look at where they are to be located. First, every installation must have means of isolation at the main intake position. Most domestic installations have a single isolator that cuts off the supply to the whole installation as shown in Figure 1.18.

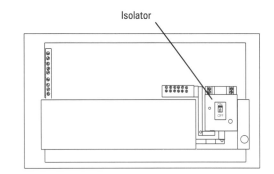

Figure 1.18   A basic domestic consumer unit

However, we do not always need to isolate the whole installation; often only specific parts need to be switched off. A cooker, for example, needs to be isolated when being repaired. The isolator, in this case, should be readily accessible, even if it is a split unit as shown in Figure 1.19.

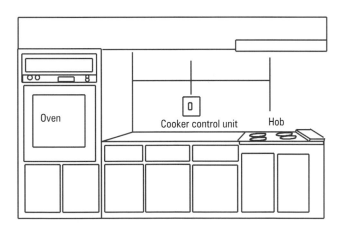

Figure 1.19   A cooker control unit should always be readily accessible.

Other examples of where isolators are local to the equipment are immersion heaters and boilers, as shown in Figures 1.20 and 1.21 respectively.

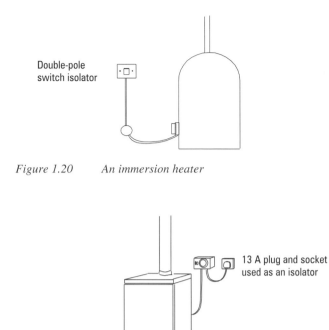

Double-pole
switch isolator

Figure 1.20   An immersion heater

13 A plug and socket
used as an isolator

Figure 1.21   A domestic boiler

In industrial installations isolators are used to control distribution boards which contain a number of circuits (Figure 1.22). Many of these circuits will supply equipment which, in turn, will have their own means of isolation.

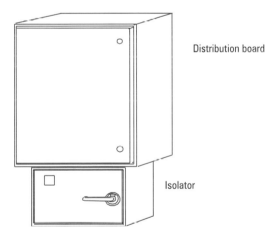

Distribution board

Isolator

Figure 1.22   A distribution board

Switching off for mechanical maintenance

Almost all electrical equipment needs to be maintained from time to time. So that this can be carried out safely, a switch should be used to disconnect the live conductors. A motor is a typical example, as in Figure 1.23. Here an isolator is placed in a position close to the motor. If it were further away, a lock off arrangement would be required.

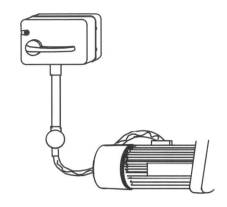

Figure 1.23   Motor isolator

Emergency switching

In workshops where there are people working on equipment which may create a hazard, emergency switches or buttons should be installed. These are designed so that if they are operated they cannot be reset until a skilled person has checked as to why they were tripped.

Another example of emergency switching is where neon signs are used for display lighting. Outside, usually below or just to one side of the sign, is a Fireman's Switch (Figure 1.24). This is installed so that, in the case of a fire, firemen are not expected to spray water over the high voltages used on the sign and expose themselves to dangerous potentials.

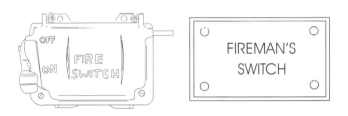

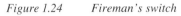

Figure 1.24   Fireman's switch

Isolation procedure

Before working on equipment check that it has been isolated and that it is dead. Following a procedure such as that in Figure 1.25 helps to ensure everything is checked.

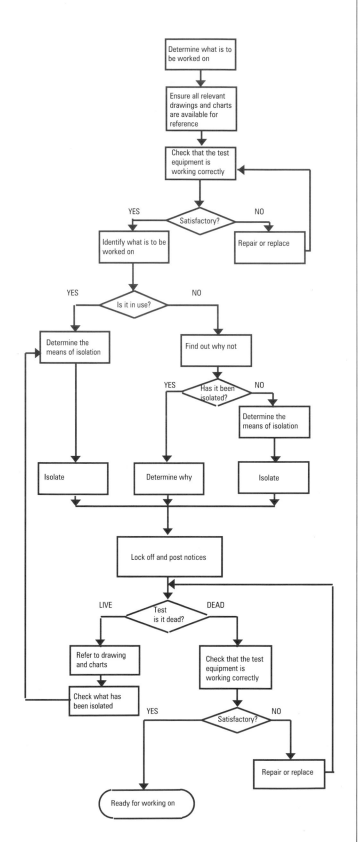

Figure 1.25   Isolation flow chart

Automatic protection against overcurrent

Overcurrent

As approximately 25% of all investigated fires are attributed to electrical faults it would seem that we should have every right to be concerned with all aspects of electrical safety. Figure 1.26 shows one type of overload that is often found.

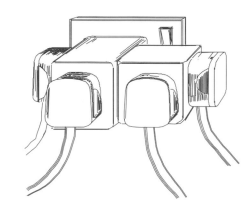

Figure 1.26

The socket outlet shown is designed to safely carry a specified value of current and if this value is exceeded the socket is said to be overloaded.

Overcurrent is where too much current is drawn through an electrical circuit. This means that some event has caused a current greater than the intended (or design) current to flow through the conductors and components of a circuit. This may be an **OVERLOAD** as the result of additional equipment being added to the circuit causing the current required to rise and exceed the design current of an otherwise healthy circuit. Alternatively an overcurrent may be a **SHORT CIRCUIT CURRENT**, which is the result of a fault within the circuit or components which results in the circuit impedance being considerably reduced. The principal difference between these two is that:

> **OVERLOAD CURRENT** may build up gradually, often over hours or days, and may remain at a level slightly above the design current for some time.

> **SHORT CIRCUIT CURRENT** tends to occur very rapidly and often reaches values considerably above the design current of the circuit.

The level of overcurrent and the characteristics of the devices to protect against the different types of overcurrent vary and so we shall look at each overcurrent in turn, beginning with the Overload.

Overload

An overload is a situation that occurs in a circuit which is still electrically sound. It is generally caused by trying to take more power from a circuit than it is designed for. This results in a larger than normal current flowing in the circuit. If the load is reduced then the circuit can continue to function without any need for repair. Each appliance connected to the socket, in the above example, is perfectly healthy and so is the circuit supplying the outlet, but more current is being drawn through the cable than was originally intended.

Every cable has some resistance and the result of drawing more current through the cable results in more heat being produced in the cable. This rise in temperature will, over a period of time, result in the insulation becoming less effective and eventually breaking down. In the case of severe overload the insulation becomes so hot it begins to melt and may even catch fire. This is obviously a serious fire risk and steps must be taken to avoid this happening.

Diversity

It is not always necessary to supply all the electrical equipment within an electrical installation with full load current at the same moment in time. Indeed it is often quite unlikely that all the equipment will be loaded to a maximum at one time. The designer of an electrical installation may make some allowances for this fact when calculating the load for the installation, this allowance is known as DIVERSITY. The application of diversity to installations requires some considerable knowledge and understanding of the intended use of the installation and the type of equipment being used. To enable us to consider the application of diversity we shall take a typical domestic electric cooker to illustrate the principal. Other final circuits supplying, for example, lighting in a domestic installation, have other allowances for diversity. These can be found in Guidance Note 1, published by the I.E.E.

As we have seen it is possible to have loads which are potentially greater than the circuit to which they are connected and still be within the safety requirements. So if we consider a domestic electric cooker (Figure 1.27) with a total load of say 15 kW, the chance of the whole load being switched on at the same time is somewhat remote. The total load is made up of a number of small loads all fitted with their own control. The controls are usually some form of temperature adjustment

which allows the heating elements to switch on and off to maintain a temperature.

The separate loads on our cooker could be:

> 2 hob elements at 2.5 kW each
> 2 hob elements at 2.0 kW each
> 1 grill at 2.5 kW
> 1 oven at 3.5 kW

The requirements for cooking generally dictate that different parts of the cooker are switched on at different times and simmerstats and thermostats are constantly switching parts of the load on and off. So even if all the loads were switched on the temperature controls would have them switching on and off and the actual load would be less than the maximum.

For domestic situations a calculation has been worked out which allows for this diversity of load on cookers. It means that when the maximum load has been calculated the first 10 A must be taken as being always there, then 30% of the remaining current should be added to the 10 A.

So for the 15 kW cooker the maximum current is

$$I = \frac{P}{U}$$

$$= \frac{15000}{230}$$

$$= 65.2 \text{ A}$$

From the 65.2 A we take the first 10 A which leaves 55.2 A of which we take 30%.

$$55.2 \times \frac{30}{100} = 16.56 \text{ A}$$

So the total assumed current is now:

$$10 + 16.56 = 26.56 \text{ A (instead of 65.2 A)}$$

Obviously a cable for the new assumed load is much smaller and cheaper than one for the maximum load. It is possible to apply diversity factors to other loads apart from cookers but care must always be taken for if it is possible to have the maximum load connected for any length of time an overload situation would occur.

Figure 1.27

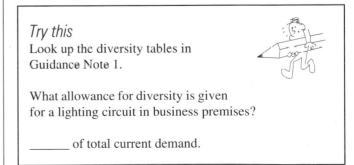

Try this
Look up the diversity tables in Guidance Note 1.

What allowance for diversity is given for a lighting circuit in business premises?

_____ of total current demand.

Overload protection devices

The usual way of protecting each cable is by installing a device in each circuit that will automatically disconnect it from the supply when an overload occurs. There are several different protection devices that can be used.

These devices include the cartridge fuses, BS 1361 and BS 1362 (Figure 1.28):

Figure 1.28   BS1361 and BS 1362

including high breaking capacity fuses, BSEN 60269-1:1994 (BS 88 part 2) (Figure 1.29):

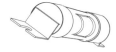

Figure 1.29   BSEN 60269-1:1994

miniature circuit breakers, BSEN 60898 (BS 3871) (Figure 1.30):

Figure 1.30   BSEN 60898

and BS 3036 fuses (Figure 1.31):

Figure 1.31   BS 3036

These will be referred to in greater detail in the next section.

Overloads are sometimes difficult for a device to detect as they can build up over a period of time as different pieces of equipment are switched on. For example, a lighting circuit is usually designed for about ten 100W lamps. If each 100W lamp was replaced by a 150W lamp and they were all switched on one after the other until they were all on, the circuit would eventually be carrying one and a half times the load it was designed for. It would depend on the characteristic of the protection device as to whether the circuit would be automatically disconnected under these conditions.

Some circuit control equipment, such as motors, have particular overload devices installed, these may be thermal, magnetic or a combination of both. These are generally installed to protect particular items of equipment, rather than fixed wiring.

14

Part 4

Short circuit

A short circuit is said to be a connection of two live conductors. "Live conductors" means all those carrying current under normal conditions, which includes the neutral conductor.

So a short circuit can occur between:
* conductors connected to different phases for example, red phase to yellow phase; red phase to blue phase, and so on
* any phase and neutral

If a short circuit is two conductors touching then it can be assumed that the resistance or impedance of that connection would be so low that it could be neglected.

> **Remember**
> The neutral is a "live" conductor.
>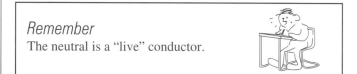

Let us consider for a moment the implications of this. If the fault has negligible impedance then the only restriction to the amount of current that will flow in the circuit is that of all the conductors.

As conductors are of a low resistance then this total value will itself be low and the current that flows can be very high. If we look at Figure 1.32 we can see the impedance of the total circuit under these conditions is made up of

the supply transformer windings (0.01 Ω)

the supply cable both phase and neutral (0.01Ω + 0.01 Ω)

the cables up to the short circuit (0.15 Ω + 0.15 Ω)

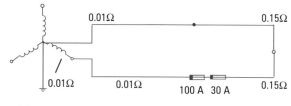

Figure 1.32

This gives us a total impedance of:

$$0.01 + 0.01 + 0.01 + 0.15 + 0.15 = 0.33 \text{ ohms}$$

If the system is operating at 230 volts then the current flow under these fault conditions will be 230 volts divided by 0.33 ohms

$$\frac{230 \text{ V}}{0.33 \ \Omega} = 696.97 \text{ amperes or } 0.69697 \text{ kA}$$

If you consider that this circuit may be protected by a 32 A device this is a large current to flow in a domestic installation.

Not only is the current flow going to be high it also tends to occur very quickly and so the protective device that we install has to be able to cope with a very different set of conditions to that of an overload. In this case the temperature builds up very rapidly, fractions of a second, and the device must sense this current and disconnect it from the supply before any damage is done to the cables or equipment. The risk of fire under these conditions is considerable.

Now if a fault should develop at the intake position of an installation the limiting impedance is only that of the supply conductors. Taking the example in Figure 1.33 the external impedance to the installation would be

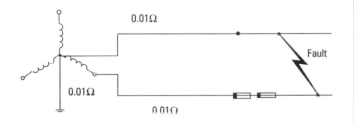

Figure 1.33

$$0.01 + 0.01 + 0.01 = 0.03 \text{ ohms}$$

this means that on a 230 V supply the fault current would be

$$\frac{230 \text{ V}}{0.03 \ \Omega} = 7667 \text{ amperes or } 7.667 \text{ kA}$$

This value is known as the prospective short circuit current at the origin of the installation.

All protective devices fitted at this point would need to be capable of breaking this current without damage to the associated equipment.

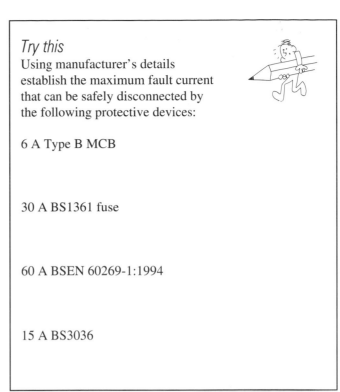

Discrimination

In any installation it is not possible to protect all circuits and equipment using a single device. Devices designed to give protection to a circuit with a load of 5 A will not serve for a circuit rated at 30 A and vice versa. A typical arrangement of an internal system to cope with this situation is shown in Figure 1.34 and we can see that there are a number of fuses between the supply intake cable and the final load of the table lamp on the socket outlet circuit.

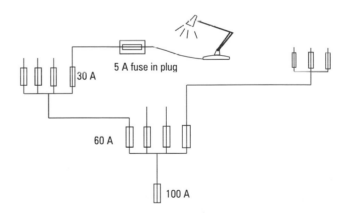

Figure 1.34   Typical arrangement providing discrimination

The objective of discrimination is to make sure that an overcurrent occurring at any point on the system causes the minimum disruption of supply. To do this we must ensure that the fuse closest to the cause of overcurrent on the supply side operates first and leaves the other devices intact thus minimising the number of circuits or appliances affected. So should a fault occur on our table lamp the plug top fuse should disconnect just the lamp from the supply leaving everything else functioning normally.

Let us now look at two examples of discrimination.

Example 1

In Figure 1.35 we have a 45 A fuse supplying a distribution board, this supplies three circuits, one of 45 A, one of 15 A and one of 5 A. If a fault should occur on the 45 A circuit then the fuse supplying the distribution board may operate and all three final circuits would then be disconnected from the supply.

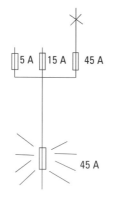

Figure 1.35

Example 2

Figure 1.36 shows a similar arrangement only this time the device protecting the board is 60 A, with the supply cables suitably rated for this. Should a similar fault occur on the 45 A circuit it would be the 45 A fuse which operates, leaving the other two circuits unaffected.

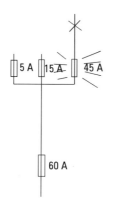

Figure 1.36

In practice the loading of all circuits should be considered before the installation begins so that the correct rating and type of protective devices can be installed. To change the protective device for one of another type or rating after completion may require extensive alterations due to the conductor sizes required.

Remember
The protection device on the supply side closest to the fault is the one that should operate.

Try this
If a fault occurs at the point shown in Figure 1.37, discrimination is said to have taken place if which protective device operates?

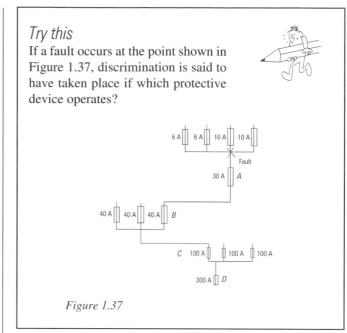

Figure 1.37

Earth leakage currents
We must also provide protection against earth fault currents.

These currents may occur due to damage to cables or equipment which result in current returning to the supply via the earth path. This includes the casing of appliances, as shown in Figure 1.38, or exposed metalwork within a building. If, while these parts are carrying current, a person or animal comes into contact with them then there is a real risk of electric shock. It is therefore important that these fault currents are disconnected as quickly as possible and for this reason regulations state that the disconnection of "Overload" and "Earth Fault" currents must be "Automatic".

Figure 1.38　　*This can lead to a dangerous situation, see the circuit shown in Figure 1.39*

The fork completes the circuit between the phase conductor and the protective conductors.

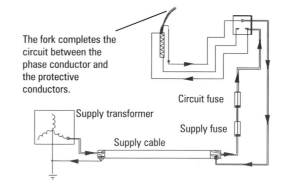

Figure 1.39　　*This shows the path that a fault current would follow in the case of a phase to protective circuit fault.*

Points to remember

Isolation

An isolator must _____ _____ an electrical installation from _____ _____ of _____.

Complete the names of the following types of isolator:

SP   – single-pole
DP   –
TP   –
TP & N   –

Overcurrent

Overload current is when _____ _____ _____ is drawn through an electrical circuit. Overload current may build up _____.

Short circuit current is a result of a _____ within the circuit or _____. They tend to occur very _____.

A short circuit can occur between:

1.
2.
3.

Diversity

For domestic situations the diversity of load on cookers means that when the maximum load has been calculated the first _____ A must be taken as always being there and ____ % of the remaining current should be added to it.

Discrimination

The _____ _____ closest to the fault on the _____ side is the one that should operate.

Self-assessment multi-choice questions

Circle the correct answers in the grid below.

1. One reason for high voltage transmission is to
   a. reduce the length of cable run
   b. keep the conductor cross-sectional area to a minimum
   c. allow more cable to be buried underground
   d. keep the conductor insulation to a minimum

2. To keep within the legal limits an electrical supplier would have to keep a stated 230 V supply between
   a. 239 and 241 V
   b. 246 and 234 V
   c. 253 and 216 V
   d. 200 and 250 V

3. A 4 wire output from a pole-mounted transformer outside a farm is most likely to be
   a. two single-phase and neutral supplies
   b. a three-phase and neutral supply
   c. a single-phase and earth supply
   d. a two-phase neutral and earth supply

4. The maximum frequency the electricity supplier can legally supply is
   a. 56 Hz
   b. 52.5 Hz
   c. 51 Hz
   d. 50.5 Hz

5. A supply cable delivering a load of 111 kW at 230 V would need to have a current carrying capacity of at least
   a. 0.4625 A
   b. 2.162 A
   c. 482.6 A
   d. 26640 A

6. On a star connected transformer winding when the line voltage is 380 V the phase voltage will be
   a. 190 V
   b. 220 V
   c. 240 V
   d. 660 V

7. The current in the neutral conductor when all three phases are carrying 20 A is
   a. 0 A
   b. 20 A
   c. 40 A
   d. 60 A

8. An installation has an external impedance of 0.4 Ω and a 230 V supply. The maximum prospective short circuit current is
   a. 173.9 A
   b. 575 A
   c. 230.4 A
   d. 920 A

9. The type of isolation on the live conductors at the intake of a domestic installation is
   a. single-pole
   b. double-pole
   c. triple-pole
   d. triple-pole and neutral

10. A means of isolation must be provided for safety during
    a. inspection
    b. operation of equipment
    c. mechanical maintenance
    d. overload

Answer grid

	a	b	c	d			a	b	c	d
1	a	b	c	d		6	a	b	c	d
2	a	b	c	d		7	a	b	c	d
3	a	b	c	d		8	a	b	c	d
4	a	b	c	d		9	a	b	c	d
5	a	b	c	d		10	a	b	c	d

Self-assessment short answer questions

1. Explain what is meant by each of the following:
   (i) Overload current
   (ii) Short circuit current

3. A domestic flat requires a new main cable to supply it. As this is part of a larger installation, it is the responsibility of the consumer to supply the cable. The loads within the flat are:

Lighting	8 × 100 W filament lamps
	2 × 80 W fluorescent lamps
Power	9 × 13 A socket outlets connected to a single 30 A ring circuit
Cooker	18 kW rated current
Water heater (Instantaneous)	1 × 3 kW

   Calculate the assumed current demand for the flat which is supplied with 230 V single phase.

   State what options may be considered by the public electricity supplier when delivering this load for sizing the cable.

2. When working through the recognised procedure for isolation explain why it may be useful to have the relevant charts and drawings.

2

Overcurrent and Earth Fault Protection

Check that you can remember the following facts from the previous chapter.

1. In star connected windings the line voltages are as follows. Calculate the phase voltages

line voltage	phase voltage
240	
425	
300	

2. For delta connected windings calculate the phase currents if the line currents are:

line current	phase current
100 A	
240 A	
2000 A	

3. Explain what is meant by "direct contact".

4. Give 3 reasons why an isolator is installed.

5. Where should isolators be located?

On completion of this chapter you should be able to:

◆ describe devices that can be used for overcurrent protection
◆ determine disconnection times from fuse and circuit breaker characteristics
◆ identify the systems of earthing known as TT, TN-S, TN-C-S and TN-C
◆ state reasons for using residual current devices
◆ explain the operation of residual current devices

Part 1

Fuses

In Chapter 1 the three main requirements for safety were listed. Having considered isolation, we must now examine both overcurrent protection and earth fault protection in more detail.

Our first type of device is the fuse (Figure 2.1) and this was once the most common type of overcurrent protection device in use. These devices have been around for more than 100 years providing protection against all types of overcurrents.

In basic terms fuses consist of a small diameter wire installed in the circuit. If an excessive amount of current begins to flow in the circuit this piece of wire gets hot. As the current flow increases it gets hotter and hotter until it finally melts and opens the circuit, and we say the fuse has "blown".

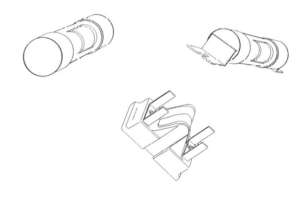

Figure 2.1   Common types of fuse

Not all fuses are made in the same way so let's look at the various types of fuses in use.

Semi-enclosed fuse (BS 3036)

This is the good old "rewireable" fuse (Figure 2.2) and there are thousands of these installed throughout the world.

It is known as a semi-enclosed fuse because the fuse element is only partially enclosed between the carrier and the base. When the carrier is removed, the fuse wire may be easily seen, and the wire element can be replaced when necessary.

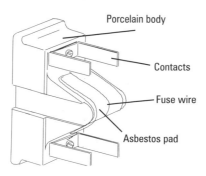

Figure 2.2   Semi-enclosed fuse BS 3036

The main **advantages** of this type are that:
- they are relatively cheap
- they are easily repaired
- they are fairly reliable
- it is easy to store spare wire
- it is easy to see when a fuse has blown

These are just some of the reasons why this type of fuse was once the most widely used overcurrent protection device. However there are some disadvantages, the effects of which have been to reduce the use of this fuse in favour of other types of device.

The main **disadvantages** are that:
- they are easily abused, the wrong size of fuse wire being fitted accidentally or intentionally
- they have a high fusing factor, around 2.5 their rated current, and as a result the cables they protect must have a larger current carrying capacity
- the precise conditions for operation cannot be easily predicted
- they do not cope well with high short circuit currents
- the wire can deteriorate over a period of time.

Cartridge fuse (BS 1361)

This fuse uses the same principle of a single fuse wire but this time the wire is enclosed in a ceramic or glass body, as shown in Figure 2.3.

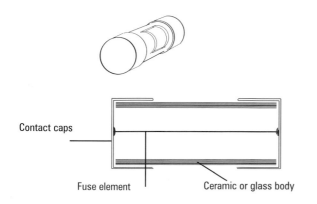

Contact caps

Fuse element   Ceramic or glass body

Figure 2.3   Cartridge fuse BS 1361

Because it is enclosed the behaviour of the fuse element under overcurrent conditions can be more accurately predicted.

Main **advantages** of this type are that:
- they have a lower fusing factor, around 1.5
- they are less prone to abuse
- being totally enclosed the element does not "scatter" when it fuses
- they are fairly cheap
- they are easy to replace
- they cope better with short circuit currents.

Main **disadvantages** are that:
- they are more expensive than BS 3036
- it is not easy to see if the fuse has blown
- stocks of spare cartridges need to be kept

This type is a reasonably cheap, more predictable, alternative to the rewireable fuse.

When the filament vaporises, the scattering metal particles are contained within the ceramic body and so present far less of a fire risk than the BS 3036 type.

Try this
Using a manufacturer's catalogue, list the ratings of the BS 1361 fuses available.

High breaking capacity (HBC) Fuse
BSEN 60269-1:1994 (BS 88 parts 1, 2 & 6)

This fuse (Figure 2.4) is the top of the range with a more sophisticated construction, which makes its operation far more predictable.

The use of a number of silver strips for the fuse element shaped as shown in Figure 2.5 means each individual strip can have a low current carrying capacity. This in turn gives the fuse a far more accurate operation. Once overcurrent occurs, the first element to "blow" increases the current flow through the others and so they operate rapidly in an avalanche effect.

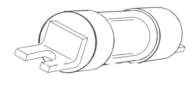

Figure 2.4   HBC fuse

The air space within the fuse body is filled with silica sand. This silica sand filler falls into the gap created by the melting elements and extinguishes the arc that is produced. This type of fuse can break short circuit currents, in the order of 80 000 A (80 kA).

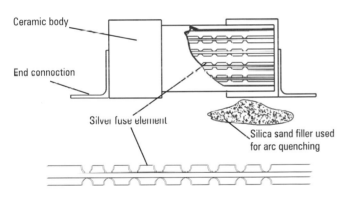

Figure 2.5

The main **advantages** of this type of fuse are that they
- have a low fusing factor, often less than 1.3
- have an ability to break high currents
- are reliable
- are accurate

The main **disadvantages** of the BS 88 part 2 are that
- they are expensive
- stocks of these as spares are costly and take up space
- care must be taken to replace them with not only the same rating of fuse but with one having the same characteristics

Circuit breakers

Instead of using fuses for protection against overcurrent we can use devices known as circuit breakers. These come in two main categories for internal use:
- miniature circuit breakers (MCBs)
- moulded case circuit breakers (MCCBs)

These devices employ a set of contacts which are automatically opened when an overcurrent occurs. This is achieved by using thermal trips for overload and magnetic trips for short circuit. Some devices are available that use only one or the other of these but most now employ both.

Circuit breakers BSEN 60898
(BS 3871)

These are used in many domestic and commercial installations.

They have the **advantages** of
- only needing to be reset after operation so no stock of replacements is required
- the setting cannot usually be adjusted
- discrimination between harmless transient overloads and yet can still cater for short circuit faults.
- it being easy to identify the breaker that has tripped.

Typical MCBs are shown in Figure 2.6.

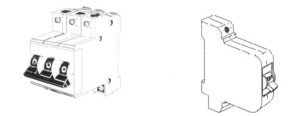

Figure 2.6   Miniature circuit breakers

The main **disadvantages** of these are that they
- are expensive
- are mechanical; physically opening the switch to break the current flow
- cannot be used if the short circuit current exceeds their short circuit rating (BS7671 434-03-01 provides an exception to this rule)

Using a mechanical switch to open circuits also creates an arc. For smaller MCBs the gap created between the contacts is sufficient to ensure that any arc is extinguished. When operating at higher current levels, and where high short circuit currents may be encountered, the circuit breaker must include some method of extinguishing the arc created. Arc splitters are fitted for this purpose.

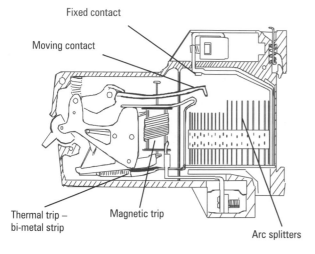

Figure 2.7   *Section across a typical miniature circuit breaker*

Characteristics

In addition to the ability of a protective device to disconnect in the event of an overcurrent, we also need to know how the device will perform under certain conditions. Each protective device has a particular characteristic and whilst these are generally covered by the particular standard to which they are produced, each manufacturer may have slightly different characteristics. The details of device characteristics given in BS 7671 are typical values for the type and precise details should be obtained from the manufacturer for the particular device being used.

The time it takes for a device to disconnect in the event of an overcurrent will depend upon two main factors:
- the characteristics of the device
- the current flow through the device

For the purpose of this exercise we shall consider the typical characteristics of the types of device used to provide protection against overcurrent. These are generally provided in the form of a graph showing the relationship between operating time and current flow and are referred as Time/Current characteristics.

The characteristics of circuit breakers to BSEN 60898 are worthy of particular mention because within the standard are a number of "types" of breaker and these types identify particular characteristics for the breakers. As these "types" perform differently, it is important that the appropriate "type" is used if the device is to provide protection against overcurrent and electric shock. For example a 16 A type "B" circuit breaker to BS EN 60898 may require a current of 80 A to cause operation within 0.1 of a second, while a type "D" may require 320 A to achieve the same disconnection time.

Earlier breakers to BS 3871 were type identified by number, for example, 1, 2, 3 and 4 and these have similar characteristic differences with the higher numeric type requiring higher currents to achieve similar disconnection times to lower numeric types.

Some applications may require a device which is less sensitive to short time overcurrents of quite high levels, where high initial starting currents occur for example. The devices used for these applications place a particularly onerous requirement on the values of earth fault loop impedance if they are also to provide protection for electric shock, as we shall see later.

Try this
Find two applications for BSEN 60898 type "B" and "D" breakers and list them below.

Type B

Type D

Examples of Time/Current characteristics are given in Figures 2.8 and 2.9 and the circuit breaker chart indicates the area of operation affected by the thermal and the magnet components of the device. Overload currents are generally dealt with by the thermal operation and fault currents by the magnetic operation.

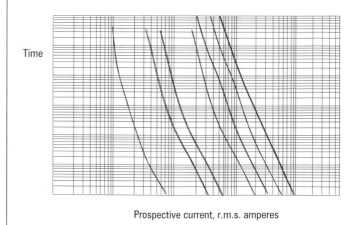

Prospective current, r.m.s. amperes

Figure 2.8   *Time/current characteristics for fuses to BSEN 60269-1:1994*

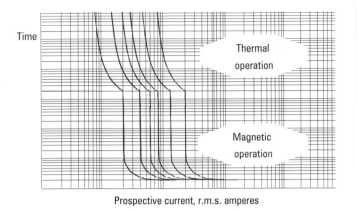

Time

Thermal operation

Magnetic operation

Prospective current, r.m.s. amperes

Figure 2.9   Typical characteristics for miniature circuit breakers to BSEN 60898

The current is shown along the horizontal axis and the time is on the vertical axis. Both of these are on a logarithmic scale and you can see that the type of device will affect the shape of the graph. We may use these characteristics to determine the speed at which a particular type and rating of device will operate for any given overcurrent, and vice versa.

BS 7671:1992 gives disconnection times for earth faults and data to enable calculations for disconnection times for overcurrents to be made.

The characteristics and operation of these devices will determine the level of protection provided and hence their ability to protect against damage or fire from overcurrents.

Example
A 10 A BS88 fuse has a prospective fault current of 65 A. The disconnection time will be found from Figure 2.10 and is 0.05 seconds or 50 milliseconds.

Try this
Using the characteristics as in Figure 2.10 below have a try at this.

A 32 A BS 88 fuse has a prospective fault current of 400 A. What will be the time for it to disconnect the supply?

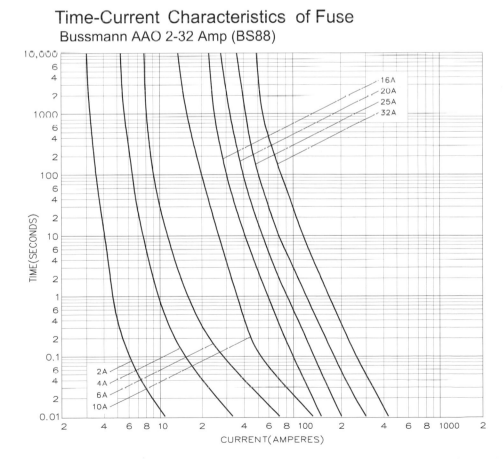

Time-Current Characteristics of Fuse
Bussmann AAO 2-32 Amp (BS88)

Figure 2.10   Characteristics for fuses to BS 88. Reproduced with kind permission of Bussmann Division, Cooper (U.K.) Ltd.

Part 2

Having considered the most common varieties of protective device, there are some circuit breakers, primarily used on the supply network and in industrial installations, which use particular methods of arc control. We shall consider the most common methods of arc control for use in particular applications before we continue.

Oil filled

In this instance the principle of arc control incorporates immersion of the contacts in oil. The contacts are usually contained in an explosion pot, as shown in Figure 2.11, constructed to aid the extinguishing of the arc. When the contacts are opened, an arc is drawn through the oil, an insulator, and gases are produced, with a little oil being carbonised in this process.

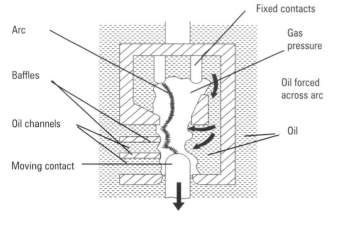

Figure 2.11   Circuit breaker explosion pot action

The pressure caused by the gas is used to force the oil around within the explosion pot, and across the arc, helping to extinguish it, the oil prevents the arc and its by-products from entering the atmosphere. In some breakers the oil is pumped across the contacts to extinguish the arc rather than rely on the pressure of the gas produced.

Remember

The oil used in transformers and circuit breakers, although contained in flameproof enclosures, is a fire and health hazard. Extreme care should be taken with the handling of oil during maintenance.

Air blast

Figure 2.12 shows the much simplified layout of an air-blast circuit breaker. As the contacts open a blast of compressed air, at around 20 bar, is forced across the arc rapidly extending its length and this quickly extinguishes it. The precise design of these breakers does vary, dependent upon the manufacturer and operational requirements. Some, for example, use a spring mechanism to open the contacts whilst others use the compressed air to open the contacts and extinguish the arc.

Obviously there is considerably more equipment required to operate this type of breaker. Air reservoirs, compressors and operating control gear and the like makes this type more complex and therefore attracting a higher initial cost. On the plus side, however, the arcing time is shorter, maintenance is cleaner and easier and the fire risk is negligible.

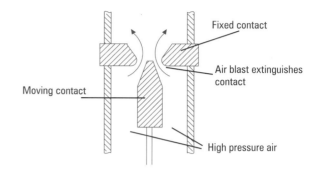

Figure 2.12   Air-blast circuit breaker

Gas

The operation of this circuit breaker is very similar to that of the air-blast circuit breaker except that a gas, such as carbon hexafluoride, is used. This is a better insulator than air or oil, is non toxic and non-flammable, inert and stable. The gas, at around 4 bar, surrounds the contacts, as they open a blast of gas at around 16 bar is forced across the arc.

Certain precautions are necessary with this type of breaker such as the installation of heaters in certain locations to prevent the gas from liquefying at low temperatures, say 9 or 10 °C. As the gas is expensive, it is usual to pump it into a storage tank during maintenance and for checks to be made regularly to detect leakage.

Vacuum

The vacuum circuit breaker is relatively maintenance free as the contacts are contained within a sealed vacuum container with the moving contact connected to the outside through a bellows arrangement as shown in Figure 2.13. The contact surfaces are in the form of flat discs and being contained in a vacuum there is no other medium to cause ionisation.

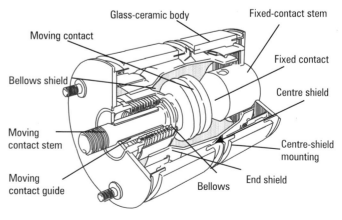

Figure 2.13   Circuit breaker explosion pot action

The result is that the arc is extinguished the first time the current passes through zero on the waveform, with minimum damage to the contact faces.

Moulded case circuit breakers

We are familiar with the construction of the MCB and the MCCB is really a more sophisticated version. The ability to break larger fault currents with this device is as a result of a more refined contact and arc control system. We can see from Figure 2.14 that one method of achieving this involves an additional set of contacts, the arcing contacts, with arc runners and an arc chute with splitter plates.

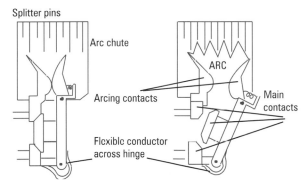

Figure 2.14   Air-blast circuit breaker

When the main contacts separate, the arcing contacts remain together and the arc is only initiated when the "arcing pair" separate. The vaporisation and heat distortion to the contacts are confined to the arcing pair. As these do not need to carry the load current during normal operation they can be made of a material such as carbon. The arc is drawn out along the route of the arc chute and the splitter plates extend the arc to create a longer run within a more confined physical space.

Alternative manufacturers' designs deal with arc control in a number of different ways but the common feature is the design of the arc chute. It is the function of the chute to increase the length of the arc as rapidly as possible over the greatest possible distance. The more able we are to do this and control the energy released the higher the fault current we can disconnect.

The balance between the thermal and magnetic operators does not really affect the fault current control as with high currents the magnetic part of the device should operate far quicker than the thermal trip, which is better placed to deal with lower overload currents. Most MCCBs have an adjustment incorporated to enable us to control the sensitivity of the device to suit various loads hence a greater versatility from a single device.

Earthing faults

Having looked at methods of protection against overcurrent, we must now consider the protection against earth fault currents.

These are generally caused by live parts coming into contact with exposed metalwork which is then made live. In order to prevent this dangerous situation from arising we connect the exposed metalwork to earth. The reason for this is to provide a safe return path for earth fault currents.

The earth fault path

There are a number of ways to provide this return path. In each case the objectives are to allow fault currents to return safely to the supply transformer and to disconnect the supply from the faulty circuit before any danger from fire, shock or burns can occur. We shall look first at the earthing arrangement for each system.

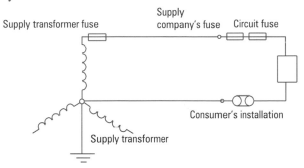

Figure 2.15

A system is a single source of supply and an installation. We shall consider each method in terms of a supply transformer, supply cables between the transformer and the installation and the installation itself. The basic circuit for this is shown in Figure 2.15.

Each type of system's earth has a particular classification which we shall use for their identification.

TT system

This tells us that:

1st letter **T** – the supply is connected directly to earth at one or more points
2nd letter **T** – the installation's exposed metalwork is connected to earth by a separate earth electrode.

The only connection between these two points is the general mass of earth as shown in Figure 2.16.

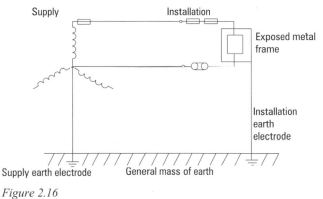

Figure 2.16

When a fault to earth occurs on this system, the earth fault current will flow around the circuit shown in Figure 2.17. This is known as the earth fault loop.

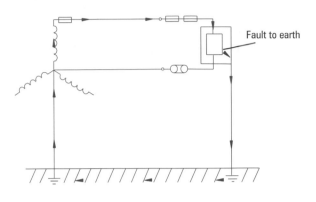

Figure 2.17

Now, in this system, the current flows through the general mass of earth which may have a very high value of impedance.

Remember

A current flow of 0.1 A, that is 100 milliamps, is sufficient to be fatal to people, so protection against even very small currents is vital to prevent danger from electric shock.

TN-S system

This tells us that:

1st letter **T** – the supply is connected directly to earth at one or more points

2nd letter **N** – the exposed metalwork of the installation is connected directly to the earthing point of the supply

3rd letter **S** – a separate conductor is used throughout the system to provide this connection (Figure 2.18).

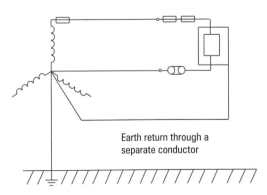

Figure 2.18

This earth connection is usually through the sheath of the supply cable and then by a separate conductor within the installation. In the event of a fault to earth in this system the current flow will be around the earth fault loop (Figure 2.19).

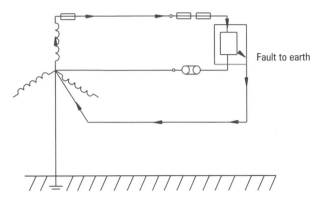

Figure 2.19

As a conductor is used throughout the whole system to provide a return path for the earth fault current, the return path should have a low value of impedance.

TN-C-S system

This type of system is similar to the TN-S system except for one important feature as we shall see.

1st letter **T** – the supply is connected directly to earth at one or more points.

2nd letter **N** – the exposed metalwork of the installation is connected directly to the earthing point of the supply.

3rd letter **C** – for some part of the system, generally in the supply section, the function of neutral conductor and earth conductor are combined in a single common conductor.

4th letter **S** – for some part of the system, generally the installation, the functions of neutral and earth are performed by separate conductors (Figure 2.20).

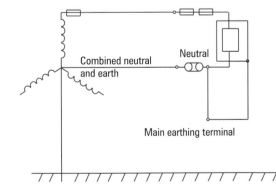

Figure 2.20

Once again the conductors are used throughout the system and so a low earth fault return path impedance should be obtained. In the event of an earth fault on this system the current flow will be around the earth fault loop as shown in Figure 2.21.

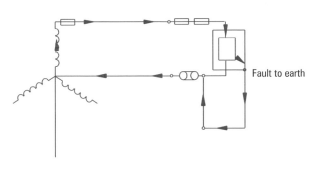

Fault to earth

Figure 2.21

These three systems are the ones in most common use, and whilst there are others we need not concern ourselves with them at this stage.

The impedance of the earth fault path plays an important part as it will regulate the amount of current that flows in the event of a fault to earth. This impedance is referred to as the earth fault loop impedance, and its value should be calculated for a proposed installation and measured for a completed installation. The measurement of the earth fault loop impedance is carried out by a special test instrument and is measured between phase and earth. We need a low value of impedance to ensure a good return path to encourage large currents to flow when a fault to earth occurs.

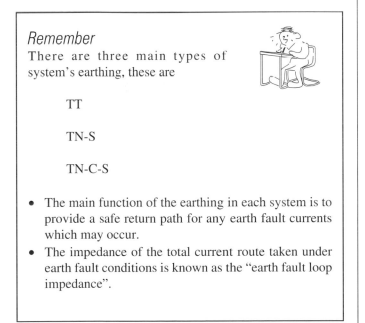

Remember

There are three main types of system's earthing, these are

TT

TN-S

TN-C-S

- The main function of the earthing in each system is to provide a safe return path for any earth fault currents which may occur.
- The impedance of the total current route taken under earth fault conditions is known as the "earth fault loop impedance".

Figures 2.22, 2.23 and 2.24 show examples of the three principal systems' earthing arrangements.

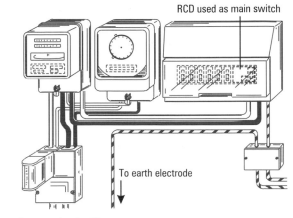

RCD used as main switch

To earth electrode

From overhead cables

Figure 2.22   TT system

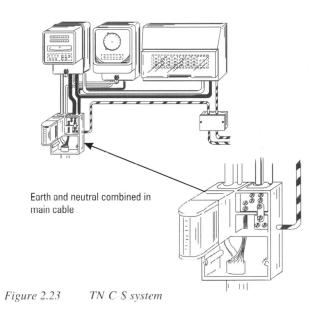

Earth and neutral combined in main cable

Figure 2.23   TN C S system

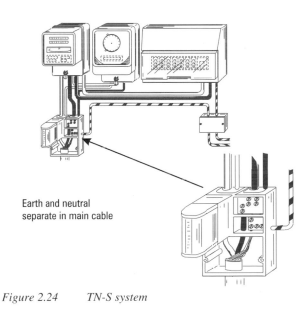

Earth and neutral separate in main cable

Figure 2.24   TN-S system

Part 3

Earth fault currents

Why do we want a high current to flow when an earth fault
occurs?

First let's consider the TN-S and TN-C-S systems and see
what happens when high currents flow to earth.

For this purpose we will assume that the fault between the
phase and frame of the equipment has no impedance at all.
Figure 2.25 shows a fault on a TN-C-S system.

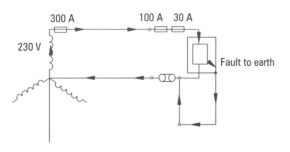

Figure 2.25   TN-C-S system

As there is no impedance in the fault itself, the only impedance
in the circuit is that of the conductors which make up the earth
fault loop. The impedance of these conductors should be low,
usually less than one ohm. This would mean that a large
current would flow.

For example, if the impedance of the conductors is 1 Ω the
fault current will be

$$I = \frac{U}{Z} = \frac{230}{1} = 230 \text{ A}$$

This would overload the cables and fuses through the system,
and as a result the "weakest" link, in this case the 30 A fuse,
would "blow" disconnecting the circuit from the supply.

We not only need to disconnect the circuit from the supply but
we must do it quickly if we are to avoid damage to cables and
equipment and provide protection against electric shock. To
ensure the current flow is high enough the earth fault loop
impedance must be kept as low as possible.

The earth fault loop impedance is made up of two principal
components, those parts of the path external to the installation
and those which form part of the consumers' installation. The
impedance external to the installation is made up from the
transformer winding in the supply transformer and the supply
company's cables, those up to the consumer's intake position.
This part of the earth fault loop path is referred to as the
"external earth fault loop impedance" and has the symbol Z_e.

That part of the loop which is formed by the consumer's
installation is again made up of two parts. The impedance of
the phase conductor up to the point at which the fault occurs,
known as R_1, and the impedance of the circuit protective
conductor(s) from the fault to the main earthing terminal,
known as R_2.

The complete earth fault loop impedance at any point on the
consumers installation, known as Z_s, is the sum of the parts
detailed above. So $Z_s = Z_e + (R_1 + R_2)$.

Remember
For a TN-S system Ze will be made up
of the transformer winding, plus the
impedance of the phase and earth
conductors.

Now let's consider the TT system (Figure 2.26).

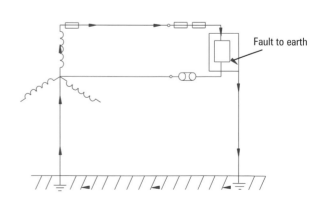

Figure 2.26   TT system

If we again assume that the fault has no impedance at all, the current flow is going to be limited by the impedance of the earth fault loop. In this case, however, the loop includes a mass of earth in place of the return conductor. The resistance of the earth itself can vary not only from area to area but also from day to day depending on the climatic conditions. It will often have a high impedance and so we cannot rely on a high current flow or a quick disconnection of the supply. In these cases we must fit a special device which detects earth fault leakage currents. The most common of these devices is the residual current device (RCD).

Residual current device (RCD)

This device senses earth fault currents by measuring current flowing into and current flowing out of an installation or circuit and compares the two. If there is a difference between the two currents the "missing current" must have returned by an alternative route, generally a current flow to earth.

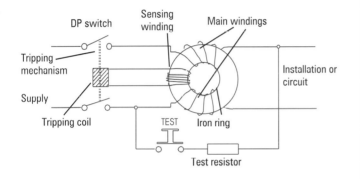

Figure 2.27

In an RCD there are two main windings wound on an iron core and these carry the phase and neutral currents. Each coil will produce a magnetic field which is directly proportional to the current that flows through it. In normal conditions these two currents will be of the same value and will consequently produce the same amount of magnetic flux.

As the two currents flow in opposite directions, the magnetic fluxes in the ring will cancel each other out and so the magnetic field in the iron ring is zero. See Figure 2.28.

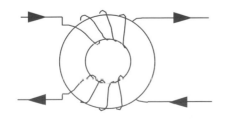

Figure 2.28

If some current flows to earth, as shown in Figure 2.29, then the phase current will be larger than the neutral current.

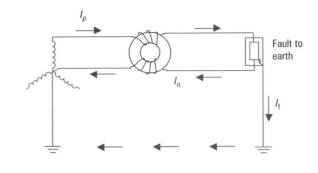

Figure 2.29 $I_p = I_n + I_f$

This means that the magnetic fluxes will no longer cancel out and a residual magnetic flux will circulate in the iron ring.

This residual flux is used to induce a small current into the sensing coil, and when this reaches a pre-set level the current flowing in this coil is sufficient to operate a small solenoid, which in turn releases the powerful spring and a tripping mechanism opens the main contacts isolating the supply. Figure 2.27

Residual current devices (RCDs) may be used to provide protection against direct or indirect contact where the conditions are particularly onerous or conventional methods cannot achieve the level of protection required. We shall consider the use of RCDs for protection against both forms of contact beginning with indirect contact.

Indirect contact protection on a TN system is generally possible using the overcurrent protection device. This is because the earth fault loop impedance is generally low enough to cause the device to operate within the time necessary to prevent a fatal electric shock. However on some installations, those on a TT system for example, the external earth fault loop impedance is too high to permit the use of this method of protection. In such cases the use of an RCD to protect the whole of the installation against indirect contact is the only alternative.

Figure 2.30

The principal of providing such protection is based on ensuring that, in the event of a fault, a potential of no more than 50 V can be sustained on the exposed and extraneous conductive parts of the installation. Remember 50 V is considered the maximum a.c. voltage at which it is unlikely that humans will experience a fatal electric shock. In order to ensure we meet this standard, the maximum operating current of the device is determined from the formula $I_a R_A \leq 50$ V, where I_a is the rating of the device causing it to operate within 5 seconds and R_A is the resistance of the earth electrode and the protective conductor(s) connecting it to the exposed conductive part(s).

Guidance Note 3 refers to the Code of Practice for Earthing (BS 7430) which suggests that an earth electrode resistance in excess of 200 Ω may prove unstable. If we considered the operating current requirements for an RCD where the electrode resistance is 200 Ω then we find that the current causing it to operate within 5 seconds must not exceed, 50 V ÷ 200 Ω = 0.25 A or 250 mA. In practice devices in the region of 100–200 mA are most common, consideration also being given to the effects of nuisance tripping causing the supply to be lost to the whole installation.

In some special locations such as construction sites or buildings housing livestock, the requirements of BS 7671 require the maximum voltage to be reduced to 25 V in the formula so the calculation becomes $I_a R_A \leq 25$ V.

Direct contact protection by use of an RCD is required in certain areas of increased risk, one typical example being a socket outlet which may reasonably be expected to supply portable equipment for use outside. For such sockets BS 7671 requires that any socket rated at 32 A or less, is protected by an RCD having an operating current of no more than 30 mA, that's 0.03 A. In addition the device should be designed to operate in 40 ms under test conditions with a residual current of 150 mA.

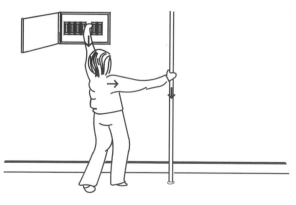

Figure 2.31

Where an installation requires an RCD to protect against indirect contact, and an RCD to protect against direct contact, for example sockets for external use, it may be necessary to include two such devices. As we have seen, these two devices have different characteristics and operational requirements. BS 7671 requires consideration be given to the inconvenience

and possible danger caused by the operation of a single protective device. If we were to fit an RCD which is suitable to provide the protection against direct contact to protect the whole installation we would normally meet the requirements for protection against indirect contact. However, the low operating current is likely to cause nuisance tripping during the normal operation of some of the electrical equipment. The effect of this would be to disconnect the entire installation from the supply on each occasion causing inconvenience and possibly endanger the user.

In such circumstances it is necessary to install two devices. We can connect these in series, but whilst they have different operating currents, in the event of a fault of high magnitude occurring, the devices are unlikely to discriminate, resulting in the whole installation being disconnected. To avoid this we can install a "time delay" RCD to provide protection against indirect contact. The use of such a device will allow us to meet all the criteria for both direct and indirect contact protection.

Alternatively we may use two devices in parallel, one for the socket outlets for outside use, rated at 30 mA, which will provide protection against both direct and indirect contact to that part of the installation. The second device will then only need to provide protection against indirect contact to the remainder of the installation and thus alleviate the need for a time delay device.

For a TT system we must ensure that the protective device is able to disconnect the supply in the event of a fault without risk of electric shock. We do this by ensuring that the voltage to earth is not allowed to rise above 50 V. This is calculated, using Ohms law formulae and the value of the earth electrode impedance, R_A, and the rated tripping current of the RCD I_a.

So we have $R_A I_a \leq 50$ V.

Where socket outlets are likely to be used to supply portable equipment for use outdoors these outlets should be protected by an RCD rated at no more than 30 mA. This requirement applies to all electrical installations, irrespective of the type of system used, ie TT, TN-S or TN-C-S. However it is not appropriate to protect the complete installation with a 30 mA RCD as the user will lose the supply to the whole of the installation in the event of the RCD operating, and at 30 mA some nuisance tripping may occur during the normal operation of the installation.

On TT installations it may therefore be necessary to install a time delay RCD to protect the whole installation and to comply with the $R_A \times I_a \leq 50$ V requirement and a 30 mA RCD to protect the socket outlets likely to be used for supplying portable equipment outdoors. Alternately two RCDs may be fitted at the origin, one to protect the socket outlets likely to be used for supplying portable equipment outdoors and the other to protect the remainder of the installation circuits.

For TN systems it is common practice to install a "split load" distribution board where some of the MCBs and hence part of the installation can be protected by the 30 mA RCD whilst the whole installation is controlled by a double pole isolator. Figure 2.32 shows a typical split load domestic consumer unit.

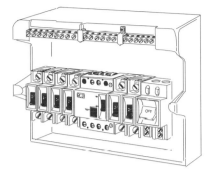

Figure 2.32   A 6-way split capacity consumer unit with 4 circuits protected by an RCD.

Internal distribution

Each of the main safety factors has now been discussed so how do they all come together in practice?

A domestic installation basically consists of three main components: the supply company intake, including their fuse and the kilowatt-hour meter, the consumer's isolation, distribution equipment and final circuits.

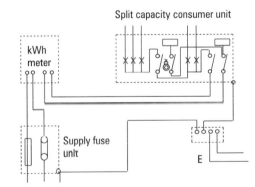

Figure 2.33   The basic circuit diagram for a split-load domestic intake.

Industrial intake and distribution systems tend to be more complex but usually follow the same pattern. Figure 2.34 shows how a number of separate units can be put together to form a more involved system.

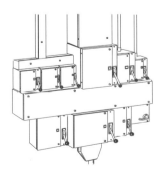

Figure 2.34   Three-phase distribution system

The circuit diagram for this looks very complex at first sight but when examined it can be broken down into separate parts. The supply is isolated by a fused switch. This is a device where the fuses are part of the switch action and move with the switching mechanism. This unit supplies a bus-bar chamber, which is basically a large junction box. From this each circuit or group of circuits is supplied via individual controls.

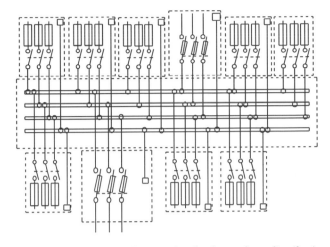

Figure 2.35   Circuit diagram for the three-phase distribution system in Figure 2.34

Each supply taken from the bus-bar chamber has its own means of isolation and overcurrent protection.

Where supplies have to be distributed throughout a building a bus-bar system in trunking is often used. This may be a vertical system with tap off points at each floor, Figure 2.36 or a horizontal system as shown in Figure 2.37.

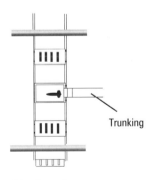

Figure 2.36   Typical busbar riser

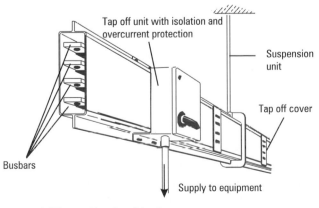

Figure 2.37   Overhead busbar system

31

In factories where a number of machines have to be supplied, an overhead bus-bar system may be used. Tap off points with conduit drops take the supply to the individual machines. The means of isolation and overcurrent protection is incorporated in the tap off unit.

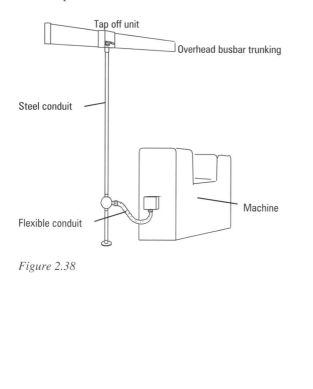

Figure 2.38

Points to remember ◀ ─ ─ ─ ─ ─ ─ ─ ─ ─ ─ ─

Overcurrent devices have to be installed throughout an installation to protect _____ and _____. There are several different types of overcurrent devices working on different principles but fundamentally they are either _____ or _____.

Fuses fit into the _____ category as they melt when the current reaches a predetermined value.

Circuit breakers often have both thermal and magnetic properties, making them suitable to detect _____ and _____ _____ currents.

Leakage currents due to earth faults must be cleared before _____ _____ or _____ can be caused. The same device as used for overload and short circuit currents can also detect _____ _____ _____. However, the circuit will need to comprise sufficiently low earth fault impedance to allow disconnection to take place within the _____ constraints. Where this cannot be achieved, and in some _____ _____, it is necessary to have a _____ _____ _____ to automatically cut off the circuit supply before harmful currents can flow.

Self-assessment multi-choice questions
Circle the correct answers in the grid below.

1. The arc quenching device for a BSEN 60269-1:1994 (BS 88) type fuse is
    a. arc chutes
    b. silica sand
    c. air blast
    d. contact separation

2. If a semi-enclosed fuse is used cables must have a higher current carrying capacity because they do not operate as quickly as other types. This means that
    a. less current flows
    b. less heat is produced in the cable
    c. no change takes place
    d. more heat is produced in the cable

3. In a BS 1361 type fuse the fuse element is contained in
    a. asbestos pads
    b. a ceramic body
    c. quartz
    d. air

4. A circuit breaker can be reset by means of
    a. a mechanical switch
    b. a blast of air
    c. an arc chute
    d. a magnetic flux

5. The two methods used to trigger the action of a circuit breaker are
    a. magnetic and hydraulic
    b. hydraulic and bimetallic
    c. magnetic and pneumatic
    d. magnetic and bimetallic

6. In a TN-C-S system the "C" tells us that
    a. the neutral and earth functions are combined throughout the whole system
    b. the neutral and earth functions are separate throughout the system
    c. the neutral and earth functions are combined in the consumer's installation
    d. the neutral and earth functions are combined for part of the system

7. The preferred method of protection against earth leakage currents in a TT system would be provided by
    a. fuses
    b. residual current devices
    c. circuit breakers
    d. isolators

8. An RCD operates when
    a. the phase and neutral currents are equal
    b. the phase and neutral currents are not equal
    c. the sensing coil current is equal to the phase current
    d. the sensing coil current is equal to the phase and neutral current

9. The maximum rating for an RCD that is to be used to protect a socket outlet circuit is
    a. 10 mA
    b. 25 mA
    c. 30 mA
    d. 100 mA

10. If a 30 mA RCD is to be installed on a 230 V supply the maximum value of earth fault loop impedance that would be acceptable is
   a. 8000 Ω
   b. 1666 Ω
   c. 8.0 Ω
   d. 1.7 Ω

Answer grid

1	a	b	c	d	6	a	b	c	d
2	a	b	c	d	7	a	b	c	d
3	a	b	c	d	8	a	b	c	d
4	a	b	c	d	9	a	b	c	d
5	a	b	c	d	10	a	b	c	d

Self-assessment short answer questions

1. On the table below list the advantages and disadvantages of the overcurrent devices named.

	Advantages	Disadvantages
Semi-enclosed Fuse (BS 3036) 1. 2. 3. 4. 5.		
HBC (BSEN 60269-1:1992) 1. 2. 3. 4. 5.		
Miniature Circuit Breakers (BSEN 60898) 1. 2. 3. 4. 5.		

2. Using the characteristics on the graph on p. 23 (Figure 2.10) determine the time it will take for a 20 A device to operate when a fault current of 300 A is flowing.

3. Explain how the Residual Current Device test button operates.

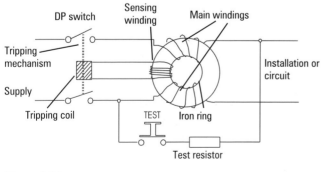

Figure 2.39

4. Sketch and label the contact compartment of an oil filled circuit breaker.

3

Electrical information required

Check that you can remember the following facts from the previous chapter.

Explain what is meant by "the fuse has blown".

Give three advantages and three disadvantages of the HBC fuse.

Circuit breakers employ a set of contacts which are automatically opened when an overcurrent occurs. Which type of trip is used for overload and which type for short circuit?

Overload:

Short circuit:

Explain what is meant by the letters in the TN-C-S system

T tells us

N tells us

C tells us

S tells us

On completion of this chapter you should be able to:

- ◆ state the information required to assess the characteristics of the supply
- ◆ list the information needed to select cables
- ◆ list the terms used in the selection process and state the meaning of each term
- ◆ state the effect of diversity on the maximum connected load
- ◆ calculate the load current for single phase circuits
- ◆ calculate the load current for balanced three phase circuits

Part 1

In the Foundation Course we looked at the effect that external influences have on the selection of a system of wiring. Now we shall establish what information we need to know about the electrical supply and expected load to enable us to select a suitable system. Once this is done, we may then begin selecting the cable sizes required to wire the installation.

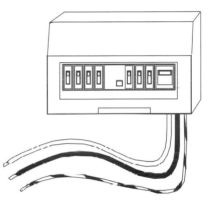

Figure 3.1   Domestic intake ready for connection

Maximum demand

The first point we need to establish is the maximum demand that will be required for the installation.

For small installations this may be done by the use of actual connected load values at an early stage in the design.

Larger installations and commercial premises are often assessed on a rule of thumb basis in the initial stages. This is because the actual loads of, for example, machines will not be known until later. Designers use a method of approximation based on the use of the building and floor areas involved. A storage warehouse, for example, would normally require a relatively low level of lighting and a minimal power load for a fairly large floor area. A small engineering unit may, by contrast, have a high level of lighting and a considerable power requirement for a much smaller floor area.

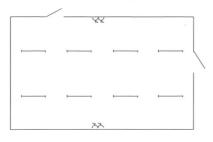

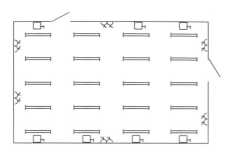

Figure 3.2   Plan of store room

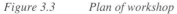

Figure 3.3   Plan of workshop

The above rooms are of the same area but the electrical load on the store room is far less than that of the workshop.

By using a table giving the load requirements per square metre for various types of building uses and the floor area to be served the designer can quickly approximate the electricity load requirements for a proposed building.

Once this is established, we would then approach the electricity supply company to determine the type of supply available. We would also need to find out if the anticipated demand can be supplied at this particular location. In doing this we ascertain from the supply company the following criteria:

The number and the types of conductors available.

In some cases, large installations and most commercial premises, we will require a three phase four wire system. In the case of small installations and the majority of domestic installations a single phase two wire will be adequate.

The type of earthing arrangement available from the company.

As it will make a difference to our installation and we need to know if it will form part of a TT, TN-C-S or a TN-S system.

In addition to this information we also need to know, from the supply company, the following details with regard to their part of the system.

The nominal voltage.

This is the voltage that we would expect to find at the point of entry to the building (the origin of the installation) and on which we will base our voltage drop constraints. This will usually be 400 V three phase and 230 V single phase but it must be verified with the relevant supply company.

The nature and frequency of the current.

If it is an a.c. supply we need to know the frequency. For the UK this will normally be 50 Hz but again it must be verified.

The prospective short circuit current at the origin of the installation.

This is the current that would flow through a fault of zero Ω impedance between phases or phase and neutral at the supply intake terminals. As this circuit comprises only the conductors normally used to supply current the impedance will be low. As a result the short circuit current will usually be high.

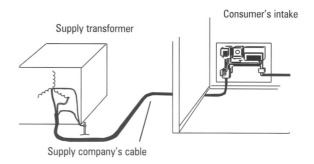

Figure 3.4   The fault current at the consumer's intake is determined by the supply transformer and cable.

The type and rating of the overcurrent protective device at the origin of the installation.

This is the size and type of fuse or circuit breaker installed by the supply company. We need to know its type and rating to ensure that discrimination occurs and that short circuit protection is provided.

Figure 3.5   BSEN 60269-1:1994 (BS 88) fuse

Figure 3.6   Cartridge fuse BS 1361

Suitability of the supply for the requirements of the installation, including maximum demand.

This is to establish that the supply is of a suitable voltage, current and frequency to supply the needs of the installation as well as providing an appropriate method of earthing.

We must also verify that the electricity supply company can supply the anticipated maximum demand of the installation. Should this not be possible we must enter into discussions with the supply company to establish the cost and necessary provision of additional equipment to achieve this need.

The earth fault impedance of the part of the system external to the installation, known as Z_e.

This is the impedance of the actual path followed by a fault current to earth in the event of a fault of zero Ω occurring between phase and earth at the origin of the installation. This is usually given by the supply authority as a **maximum expected value.**

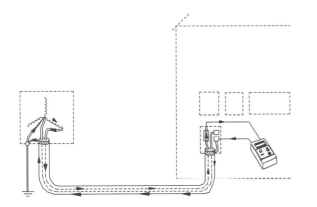

Figure 3.7   A test at the consumer's intake measures the fault current taking into account the supply transformer and cable.

If the installation has an existing supply, these values can be measured by using the appropriate test instruments. They may vary in comparison with those given by the supply company as the expected value. Generally this is relative to the location of the installation with respect to the supply company's transformer.

Once we have ascertained this information we may begin to look at the actual maximum demand requirements.

Remember
Z_e is the term used to denote the external earth fault loop impedance.

Points to remember

Before we can begin the selection of cables, we need to establish all the characteristics of the supply.
These must include: (8)

1.

2.

3.

4.

5.

6.

7.

8.

Part 2

Maximum demand

We will consider the maximum demand of an installation to be the power required if every piece of equipment installed was switched on.

Let us consider a simple example in a domestic installation with the following equipment installed:

	Total
1 × cooker rated at 45 A	45 A
2 × ring circuits rated at 30 A	60 A
1 × immersion heater rated at 15 A	15 A
2 × lighting circuits rated at 5 A	10 A
	130 A

We have established a total load of 130 A for our domestic installation but it is unlikely that the supply company will install a service larger than 100 A for this installation. This is the result of applying a diversity to our total connected load. This is a method by which an estimate of the total required load is established based on the assumption that all the equipment connected will not all be used at the same time. We shall not be considering this in detail at this time as we are concerned with the selection of cables to supply specific load currents.

Now that we know the information that we need to begin our selection process we can do a quick refresher in calculating the current requirements of given loads. Remember that for a resistive load our power formula is

power = voltage × current

If our load is inductive then we must include the power factor in this calculation and so our formula becomes

power = voltage × current × power factor

If the load connected is given in watts or VoltAmperes then by rearranging the formula we can find the current that will be drawn. For our resistive load the current drawn will be given by

$$\text{current} = \frac{\text{power}}{\text{voltage} \times \text{power factor}}$$

If we now consider a balanced three phase load then the power required for a resistive load will be given by the formula

power = $\sqrt{3} \times U_L \times I_L$

and for an inductive load

power = $\sqrt{3} \times U_L \times I_L \times$ power factor

By rearranging these formulae as before we can determine the line current required for a balanced three phase load.

$$\text{line current} = \frac{\text{power}}{\sqrt{3} \times U_L \times \text{power factor}}$$

Remember
Power factor is, in fact, the cosine of the phase angle between the voltage and current, known as cos θ, so the formula may appear as

$P = UI \cos \theta$ for single phase and so on.

Let's consider a simple example.

A circuit is to supply an electric heater at 230 V 50 Hz and this heater is rated at 3.5 kW. What will be the current drawn from the supply if the power factor is unity (1)?

Remember
Purely resistive loads will have a power factor of unity.

We usually consider a heater of this size to be a resistive load with a power factor of 1.

Power (watts)	=	Voltage (volts) × current (amps) × power factor

$$\text{Current} = \frac{3500}{230 \times 1}$$

$$= 15.217 \text{ amperes}$$

This is the "Design Current" for this heater and is given the symbol I_b. It is the current that will be drawn by this heater under normal operating conditions.

Remember
I_b is the design current of a circuit or load and is the current drawn from the supply under normal conditions.

If we now apply the same principle to a balanced three phase load of, let's say, 15 kW at 400 V, 50 Hz and a power factor of unity (1) then we have:

$$\text{Power} = \sqrt{3} \times U_L \times I_L \times \text{power factor}$$
$$15000 = \sqrt{3} \times 400 \times I_L \times 1$$

so the line current will be

$$I_L = \frac{15000}{(\sqrt{3} \times 400 \times 1)}$$

Calculate the terms inside the brackets before you divide.

$$= 21.65 \text{ amperes}$$

So for this balanced three phase load the design current I_b will be 21.65 amperes per phase.

Try this

A single phase load of 2.5 kW is to be supplied at 230 V 50 Hz. If the load has a power factor of 0.9 what will be the design current of the circuit?
(Remember to calculate the terms inside the brackets before you divide.)

Try this

We are to supply a balanced three phase load at 400 V 50 Hz. If the power required is 25 kW at power factor of 0.8 what is the design current?

In the Appendix at the end of this book there is a project which you have to complete before you can be credited with this module. To assist you, at the end of this chapter and Chapters 4, 5, 6 and 7, a part of the project has been given as an example. Turn to the Appendix now and read through the project. You will see that the cooker circuit (Circuit 1) is the circuit that we will be using as our example throughout this book.

The cooker circuit (Circuit 1 of the project) is run as shown on the drawing below. The route that has been chosen will be used for all of the example calculations at the end of this chapter and Chapters 4, 5, 6 and 7.

By following the steps used in the examples you should be able to apply the same principles to the circuits referred to in the project. Circuit 2 should be completed, using the route shown in the Appendix, at the end of this chapter and Chapters 4, 5, 6 and 7.

Using the specification and drawings shown in the Appendix the following information can be obtained.

External factors:

- The majority of these details are given in the specification. We know the type and use of the building, its construction and the fire escape routes are well defined.
- We know that the environment will be relatively dirt free and that there should be no problem with humidity or water.
- The areas with high ambient temperatures have been highlighted and are the kitchen and boiler room. To the best of our knowledge there is no danger from corrosive or polluting substances.
- We shall assume that the persons using the building are able bodied and capable.

These factors will apply to all the circuits in the Project.

The information that we need to be able to make a suitable selection of system is also given within the specification. We can best look at this as a list of data:

Number and type of conductors	3 phase 4 wire
Type of earthing arrangement	TN-C-S
Nominal voltage	400/230 V
Nature and frequency of current	50 Hz a.c.
Prospective short circuit current at the origin of the installation	16 kA
Type and rating of overcurrent device at the origin	100 A BSEN 60269-1:1994 (BS88)
Earth fault loop impedance external to the installation (Ze)	0.3 Ω

Suitability of the supply:
The supply would appear to be suitable for our needs but we must be aware of the total loading for the installation and compare this with the rating of the supply company's protective devices at the origin to ensure that this is suitable. This will be beyond the scope of this project but this check should be made for every installation as the design progresses. **Remember** that the rating of the supply will normally be requested based on an estimated value. Checks should be made to ensure that the requested value is adequate as the design progresses.

For this project we have to consider a submain from the main distribution board on the first floor. The detail for this submain is also given in the specification. When we carry out our calculations for the first floor, we must use the submain detail to ensure full compliance. The values given for the submain are those at the incoming terminals of the first floor distribution board.

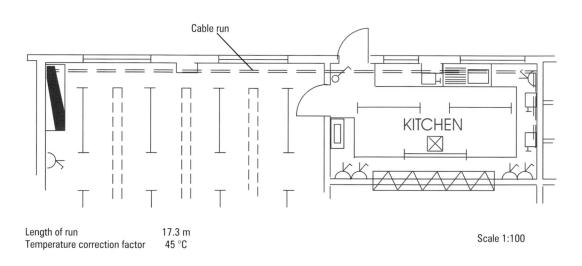

Cable run

Length of run	17.3 m
Temperature correction factor	45 °C

KITCHEN

Scale 1:100

Number and type of conductors	2 single phase
Type of earthing arrangement	TN-C-S
Nominal voltage	230 V
Nature and frequency of current	50 Hz a.c.
Prospective short circuit current	5412 A or 5.4 kA
Type and rating of overcurrent device at the origin	63A BSEN 60269-1:1994 (BS88)
Earth fault loop impedance at first floor distribution board $(Z_c) + (R_1 + R_2)$ of submain cable	0.824 Ω

Remember: we must use the details of the submain at the first floor distribution board to carry out our calculations for the first floor circuits.

Self-assessment short answer questions

1. State the two methods of determining the maximum demand for a proposed installation in the early stages of design, giving an example of the type of installation for which each would be used.

2. What is meant by the "design current" of a circuit and what symbol is given to denote it?

3. List ALL the information that you would need to obtain from the supply company to enable you to select a suitable system of wiring and size the cables required.

4. If a single phase load of 3.5 kW is to be supplied from a 220 V 50 Hz supply and has a power factor of 0.95, what will the circuit design current be?

5. Calculate the power factor for a 25 kW balanced three phase load if it is supplied at 450 V 50 Hz and has a line current of 45 A.

4

Selecting the conductor size

Check that you can remember the following facts from the previous chapter.

List the information required from the electricity supply company required to assess the characteristics of the supply.

What is the term used to denote the external earth fault loop impedance?

The formula for calculating power in a circuit containing a resistive load is:

The formula for calculating power in a circuit containing an inductive load is:

On completion of this chapter you should be able to:

◆ determine the load currents for given circuits
◆ select the correct rating of protective device for given circuits
◆ state the conditions that require the application of correction factors
◆ apply the correction factors to determine the minimum current carrying capacity for the cable to supply given loads

Part 1

In the previous chapter we established what information we needed to have before we can begin to select the wiring system in detail. We have also calculated the design currents for single and three-phase circuits. Now we shall begin the process of selecting the correct size of cable to supply the various parts of the system. This includes all the cables that go to make up the consumer's installation. For larger installations the consumer may purchase energy at 11 kV or above. In these instances it would be necessary to determine the size of the consumer's own supply cables.

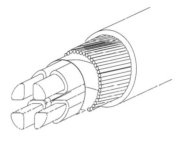

Figure 4.1   Typical distribution supply cable

Conductor selection

In some cases we may find that the installation imposes conditions that cause us to reconsider our cable sizing and choose an alternative method or system in the interests of practicality, economy or ease of installation. The first part of the process is to determine the design current of the circuit. You should remember that this is the current that the load will require under normal conditions.

Consider the simple domestic electric heater and we will assume that this has a total power requirement of 3 kW when supplied at 230 V 50 Hz.

Design current

Using the data we are given (remember that being a resistive load we will consider that the power factor is unity) we get a required current of:

Required current $= \dfrac{3000}{230 \times 1}$

$= 13.04$ amps

So if 13.04 amperes is the current required for the heater to operate normally then logically the fuse or circuit breaker used to protect the circuit must be capable of carrying this current without damage or deterioration for an indefinite time.

Now protective devices are manufactured in a standard range of sizes, so the next task is to determine the appropriate size of device to be used. The actual rating will depend on the type of device used so we must decide on the type before we can select the rating. Some of the types and ratings are shown in Table 4.1.

Table 4.1

Protection type	Current rating
BSEN 60269-1:1994 (BS 88)	2
	4
	6
	10
	16
	20
	25
	32
	40
	50
	63
	80
	100
	125
	160
	200
	250
	315
	355
	400
BSEN60269-1:1994 (BS 88)	355
	400
	450
	500
	560
	630
	710
	800
BSEN60269-1:1994 (BS 88)	2
	4
	6
	10
	16
	20
	25
	32
BS 1361	15
	20
	30
	40
	45
	50
	60
	80
MCB BSEN 60898/BS3871	Types B & C (2 and 3)
	6
	10
	16
	20
	32
	40
	50
	63

In this case we are going to use a BS 1361 type fuse and by reference to the manufacturer's data or the data in Table 4.1 we can see that the nearest size of fuse is a 15A. This is the nominal rating of the fuse, that is to say the current that the fuse will carry for an unlimited period of time without deterioration. It is referred to as I_n and it is important that we remember that the rating of the device I_n is equal to or greater than the design current for the circuit I_b .

This may be written as $I_n \geq I_b$.

Table 4.2

Fuses to BS 1361							
Rating (amperes)	5	15	20	30	45	60	80	100
Z_s (ohms)	17.1	5.22	2.93	1.92	1.0	0.73	0.52	0.38

This rule must always be applied so for example if the design current of a circuit was 20.5 A we could not use a 20 A BS1361 fuse for protection and we would have to go to a 30 A device.

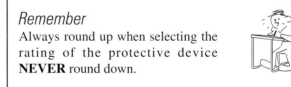
So for our domestic heater we know that we need to install a 15 A fuse. The next stage is to determine the current carrying capacity of the cable and so select a suitable size of conductor. The first thing to establish is the minimum current carrying capacity of the cable required and we shall refer to this as I_t .

However, before we can do this we must give some consideration to the conditions under which the cable will be operating and the method of installation.

Whatever type of material we use for a conductor it will have some resistance. We know that the factors, which affect the value of this resistance, are
• the material from which the conductor is made
• the cross sectional area (c.s.a.) of the conductor
• the length of the conductor
If we pass current through a resistor, we produce heat and a voltage drop occurs across the installation.

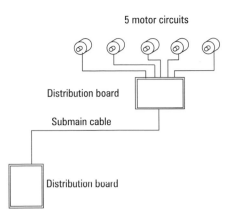

5 motor circuits

Distribution board

Submain cable

Distribution board

Figure 4.2   When all of the motor circuits are loaded the voltage drop on the submain cable should not be excessive

A rise in temperature will also produce an increase in the resistance of almost all metallic conductors. We must try to keep the heat produced in the conductor to a minimum during operation. To enable us to do this we must consider the types of location and conditions which will affect the heat that a cable can dissipate (give off).

The main points are:
- ambient temperature
- grouping
- thermal insulation
- type of protection device used

Ambient temperature

This is the temperature of the surroundings of the cable. It is often the temperature of the room or building in which the cable is installed. Now, the hotter the surroundings the less heat the cable will be able to give off. We put food in a warm oven to keep it hot. If the surrounding temperature is low then the heat given off is greater and so the cable would "run cooler", and so could give off more heat.

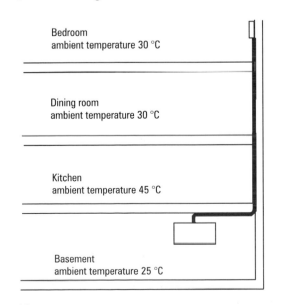

Bedroom
ambient temperature 30 °C

Dining room
ambient temperature 30 °C

Kitchen
ambient temperature 45 °C

Basement
ambient temperature 25 °C

Figure 4.3

Grouping

If a number of cables are run together, they will all produce heat when they are carrying current. The effect of this is that they reduce the heat dissipation by each other. The same effect is used by groups of animals huddling together to keep warm. If we keep the cables separated then this effect will be minimised.

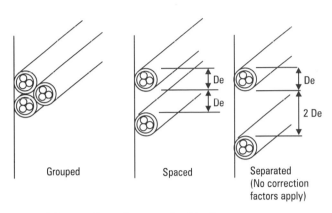

Grouped   Spaced   Separated
(No correction
factors apply)

Figure 4.4   De = the overall cable diameter

Thermal insulation

This has the effect of wrapping a cable in a fur coat on a hot summer's day and the heat produced cannot escape. In terms of electrical installations there are two conditions to be considered, these being
- those in contact with thermal insulation on one side only, such as in Method 4 in the scheduled methods of installation of cables in BS 7671:1992
- cables completely surrounded by thermal insulation

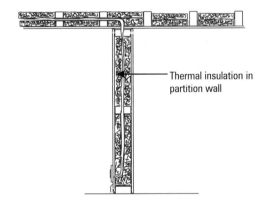

Thermal insulation in partition wall

Figure 4.5

Type of protection device used

When we use a device to protect a cable, it usually operates when too much current is drawn through it. Excess current will produce more heat within the cable and unless the device can disconnect quickly enough damage may be caused to the cable insulation. Under extreme conditions the insulation may catch fire.

The protective device that we install is to protect the _____ supplying the equipment.

Complete the following sentence:

The "nominal rating" of the fuse is the current

The factors which affect the value of the resistance of a conductor are: (3)

If current is passed through a resistor _____ is produced and a _____ _____ occurs across the resistor.

A rise in temperature will also produce a rise in _____ in almost all metallic conductors.

When selecting the rating of the protective device always _____ to the nearest size.

The ambient temperature is the temperature of the _____ of the cable.

If a number of cables are run together, they will all produce heat when they are carrying current. How can we minimise this effect?

In terms of thermal insulation in electrical installations there are two conditions to be considered. They are:

1.

2.

When we use a device to protect a cable, it usually operates when too much _____ is drawn through it. If the device does not disconnect quickly enough what damage may be caused?

Part 2

Applying factors

Now that we have seen the situations and conditions that can cause too much heat in a cable what can we do to prevent this occurring?

Whilst we cannot prevent some of these conditions, we can take some precautionary steps. Modification of the route to avoid a particular area or installing cables with adequate space between them are two examples.

This will not always be possible and so we must have some other approach to ensure a safe installation. To do this we use a system of FACTORS, one for each condition. The purpose of these is to cause us to use a larger c.s.a. of conductor to reduce the resistance and so reduce the heat generated within the cable.

Let's look at the factors we use and see how they are applied. These factors are given in BS7671:1992 so we shall also consider where they are to be found within that document.

Ambient temperature

The tables used for cable selection will be based on a particular ambient temperature. In the case of BS7671:1992 this is 30 °C so any cables installed in an ambient temperature above this will need to be adjusted as they cannot give off as much heat. A set of factors for these conditions are given in BS7671:1992 in Table 4C1 and 4C2. You will notice that the type of insulation used also has an effect on the factors given in the tables. This factor is referred to as C_a.

Try this

Complete the table below using a current copy of BS7671:1992.

Correction factors for ambient temperature

Type of insulation	Ambient temperature °C							
	30	35	40	45	50	55	60	65
60 °C rubber (flexible cables only)								
General purpose PVC								

Table 4.3

Grouping

The factors used for grouping are contained in Table 4B1. By reference to 4B1 you will see that the method of installation also has some bearing on the factor to be used. It is important to remember that these factors are applied to the number of circuits or multicore cables that are grouped and not the number of conductors, an important point when we are installing in conduit or trunking. This factor is known as C_g.

Thermal insulation

This is dealt with in two ways. First if the cable is surrounded by thermal insulation then the factor that must be applied is a constant value of 0.5 (or for cables up to 10 mm^2 over short distances as shown in Table 52A), this factor is known as C_i. If the cable to be installed is in contact on one side only with the thermal insulation then this situation is dealt with by using the tables for this installation method. We shall consider this later as we do not have to apply a factor for this particular situation.

Type of protective device used

As we noted earlier, this factor is really dependent on the speed and reliability of the protective device. It is sufficient to say that the only device that does NOT disconnect in sufficient time is the BS3036 semi-enclosed rewireable fuse.

Figure 4.6

If we select this particular device then we must **ALWAYS** use a factor of 0.725 when calculating current carrying capacity. This value is derived from Regulation 433-02-03 which states that for this type of device I_n should not exceed 0.725 × the lowest cable current carrying capacity, giving us our factor C_f for the fuse.

$$I_n \leq (0.725 \times \text{lowest cable current carrying capacity})$$

So how do we apply these factors to determine the size of conductor that we require? You will notice that nearly all of these factors are less than 1. We also know that the current carrying capacity of the cable must be equal to or greater than the rating of the fuse that protects it. The purpose of applying the factors is to ensure that the conductor is large enough to carry the current without excessive heat being generated and if the dissipation is affected by adverse conditions then the only way in which this can be resolved is by increasing the size of the cable.

When a protective device operates it generally relies on an overcurrent to do so. If a device is rated at 20 A then this is the current that it will carry for an indefinite period without deterioration. It follows then that the device will only register current beyond its rated value as overcurrent so we must use the rating of the device to carry out our calculations to find the value of I_t.

If more than one of our conditions for the application of factors exists then we must consider the worst case. If, as an example, a cable runs through an area of high ambient temperature then it is grouped with several other cables at another location and is finally run totally enclosed in thermal insulation at another point then we consider all the factors and apply the most onerous. However, if more than one condition applies at a single location then we must apply all those factors.

To establish the value of I_t for any circuit we must divide the current rating of the protective device by all the factors. If any factor does not apply, we give it the value 1.

So
$$I_t = \frac{I_n}{(C_a \times C_i \times C_g \times C_f)}$$

Where C_f is the factor for a BS3036 fuse.

If no factors need to be applied then

$$I_t = \frac{I_n}{(1 \times 1 \times 1 \times 1)}$$

and in this case

$$I_t = I_n$$

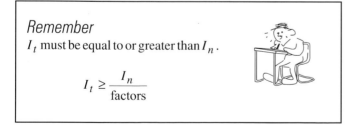

Example
If the cable supplying our heater is run through a roof space with an ambient temperature of 35 °C, and if this is the only factor that applies, then we can calculate the value of I_t by simply dividing the fuse rating by the factor for an ambient temperature of 35 °C .

So we have

$$I_t = \frac{15 \text{ A}}{C_a \times 1 \times 1 \times 1}$$

So we turn to Table 4C1 (remember we are using a BS1361 fuse). The first thing that we find is that we need to know more details about the circuit that we are to install. In this case we need to know the type of insulation that we are to use. In this example we shall use a PVC PVC cable as it is a normal domestic installation. This means we have the value of 0.94 for general purpose PVC.

So our value for I_t will be

$$I_t = \frac{15 \text{ A}}{0.94} = 15.957 \text{ A}$$

You can see that the current carrying capacity has increased to almost 16 A as a result of the effect of a higher ambient temperature.

I_t for this situation is 15.957 A

If the same cable is to be run through the roof space but is grouped with 3 other cables, in an ambient temperature of 35 °C, and it is protected by a BS3036 type fuse then we must apply all these factors to establish I_t so we get

$$It = \frac{15 \text{ A}}{C_a \times C_g \times C_f \times C_i}$$

(C_f for the BS3036 fuse is 0.725)

$$C_a = 0.97 \text{ from Table 4C2}$$

When we come to establish the value of C_g using Table 4B1 BS7671:1992 we find that we need yet more information about the circuit that we are to install. This time we need to know the method of installation. This is the method of wiring, so we need to know how the cable is to be installed. For the normal domestic installation the method would be to run the cables clipped direct to the surface of the building so that is the method that we shall use.

$$C_g = 0.65 \text{ from Table 4B1}$$

We must remember to use the total number of cables that are bunched together. For this example this is the 3 other cables plus the 1 cable that we are to install giving a total of 4.

C_i does not apply in this case as the cable is not surrounded by thermal insulation at any point on the route, so $C_i = 1$.

So to complete the calculation we have

$$I_t = \frac{15}{0.97 \times 0.65 \times 0.725 \times 1}$$

$$I_t = 32.81 \text{ A}$$

As you can see, this has a considerable effect on the current carrying capacity of the conductor required. In practice it would be a more sensible idea to run the additional cable through the roof separate from the other cables and consider an alternative protective device. This would then reduce the number of factors that need to be applied.

Try this

Using the cable from the above example we are going to install it for the heater so that it is clipped separately from all the other cables.

Calculate the value of I_t for this situation if all other factors remain unchanged.

There are going to be situations where a number of factors will need to be applied but not all at the same point along the route. To cater for these situations we take the worst condition that exists along the route and size our cable to those requirements. If it complies with the worst conditions then it must be satisfactory for the rest.

Before we can carry out this exercise we must have the following information:
- the type of protective device that is to be used
- the type of cable that is to be installed
- the method of installation to be employed
- the ambient temperature that is likely to be present along the route
- any grouping that may occur along the route
- any thermal insulation that may be encountered on the way

This will provide us with the most onerous conditions that we will encounter. We can, of course, reconsider our route, method of installation, type of cable or protective device if we can in order to produce the most economic and practical installation we can.

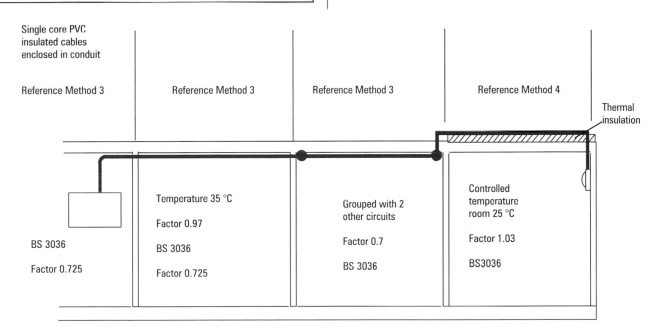

Figure 4.7   The most onerous conditions in the above diagram are where the cables are grouped with two other circuits. This gives a factor of 0.725 × 0.7 = 0.5075. (The factor for the BS 3036 fuse applies throughout the whole length of cable run.)

_____ need to be applied to make allowance for the adverse installation conditions that are encountered by the cable. If we can reduce these then cable size may be kept to a _____.

List items of information that we must obtain before we can establish the most onerous conditions on a cable route:

We may need to re-route and so on to reduce the numbers of factors that have to be applied, thus preventing the cable size becoming impractical.

The following symbols are used in BS7671. Using Appendix 4 in BS 7671 describe what these symbols mean:

I_t

I_b

I_n

C_a

C_g

C_i

C_t

PROJECT

Refer to the specification and drawings shown in the Appendix.
Example calculations use the cooker circuit (Circuit 1).

Selecting the conductor size:

Determine the design current of the circuit.

> Connected load = 8.5 kW
> Apply diversity from Table 4B.
> If we assume this is the largest cooking appliance we must allow 100% of its full value.

Design Current $I_b = \dfrac{8.5 \times 10^3}{230}$

$$= 36.956 \text{ A}$$

Select protective device rating

> $I_n \geq I_b$ from Table 41D BSEN 60269-1:1994 (BS88 Pt 2 & Pt 6) fuses = 40 A (with maximum Z_s of 1.4 Ω)

Minimum current carrying capacity

$$I_t = \frac{I_n}{C_a \times C_g \times C_f \times C_i}$$

$$= \frac{40}{0.71 \times 1 \times 1 \times 1}$$

$$= 56.34$$

Where C_a = 0.71 Table 3.3 (4C1) – ambient
temperature 45 °C
C_g = 1.0
C_f = 1.0 (correction factor for BS 3036 fuse)

Now calculate I_b, I_n and I_t for the reception area lighting circuit in the project (Circuit 2). Use the space provided on the next page.

Project – Circuit 2

1. State:
    a. The effect on the current carrying capacity of a cable that is totally enclosed in thermal insulation

    b. What effect this will have on the size of the cable required and why.

2. State the effect that grouping cables together has on their current carrying capacity and why this is so.

3. A PVC/PVC cable is to supply a load of 2 kW at 230 V, 50 Hz and unity power factor. If the circuit is to be protected by a BSEN 60269-1:1994 type fuse and the cable will be grouped with 2 other cables throughout the run what will be the value of I_t for this circuit?

5

Selecting the Cable

Check that you can remember the following facts from the previous chapter.

If the design current of a circuit is 16.5 A and the nearest BS 1361 protective devices are rated at 15 A and 20 A which would be selected?

Cables run together that are "spaced" are how far apart?

Cables run together that are "separated" are how far apart?

Which of the following is correct?

$$I_t \geq I_n$$

$$I_n \geq I_t$$

On completion of this chapter you should be able to:

- select the correct cross-sectional area conductor, to supply given loads, from BS 7671.1992
- calculate voltage drop for given circuits
- calculate the maximum permissible voltage drop for a system
- verify that circuits comply with the volt drop constraints

Part 1

In the previous chapter we calculated the minimum current carrying requirements for conductors to supply a given load. In this chapter we will consider the selection of suitably sized conductors from the tables to supply loads under defined conditions and ensure compliance with the voltage drop constraints.

Figure 5.1

The first stage is to determine the minimum current carrying capacity of the conductor.

To do this we will use a simple circuit which has no factors to be applied.

A circuit is to supply a load of 16 A at a distance of 20 m from the supply intake position. The supply is 230 V 50 Hz and the circuit is to be protected by a BS1361 type fuse. The circuit is to be installed using PVC/PVC cable clipped direct to the surface and the conditions are such that no factors need to be applied.

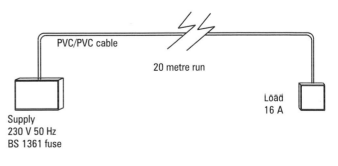

PVC/PVC cable

20 metre run

Supply
230 V 50 Hz
BS 1361 fuse

Load
16 A

Figure 5.2

The first thing we shall do is to establish the installation method. We know how the cable is to be installed and so we can find the method reference number in BS 7671:1992 Appendix 4, Table 4A. You can see that the table is divided into four columns. Column 1 gives a reference number to each method of installation and column 2 gives a written description of the method. The third column gives an illustration of the method that has been described and sometimes additional qualification of the requirements. All three columns are used to establish the way in which the cable is to be installed and once this is done we can read off the reference method from column 4.

For our example then we can see that the description in the first row of column 2 is the one that describes the way we are to install our cable so the reference method from column 4 is Method 1.

Now we can carry out the calculations to enable us to select conductor size.

$I_b = 16$ A and $I_n \geq I_b$ so from the tables we can determine that the size of BS1361 fuse to be used is 20 A. As there are no factors to be applied then we can see that

$$I_t = \frac{20}{1 \times 1 \times 1 \times 1}$$

$$= 20 \text{ A}$$

Our next step is to select the cross sectional area of the conductor that we are to use and we do this with the aid of the tables in Appendix 4 of BS 7671:1992. The cable size is selected on the basis of the current carrying capacity of the conductor under the stated conditions. This is where the method of installation is important because we use this to establish which column of the table to use.

We must select the appropriate table in Appendix 4 to find the correct size of conductor. Tables 4D1A to 4L4B contain the relevant information on current carrying capacity and voltage drop. If we look below the table heading, we find a description of the cable type, insulation and conductor material. Table 4D1A for example deals with single core PVC insulated non armoured cables with copper conductors. As we are not going to install single core cables, this table is not appropriate. We are to install PVC/PVC cables, and these will come under the heading of multicore cables. Table 4D2A covers the type we are using so this is the table we shall use.

The actual current carrying capacity of the cable is known as I_z and as you may imagine the value of I_z must be at least equal to the value of I_t for the circuit in question. So we can say that $I_z \geq I_t$.

So our cable selection is made in the following stages
1. Select the appropriate table
2. Select the vertical column that relates to the reference method that we are using
3. Select the column for the type of circuit, i.e. single or three phase
4. Move down this column until a value that is equal or greater than I_t is reached
5. Move horizontally across the columns to column 1 and read off the cross sectional area of the cable

Now, for our example, we need a cable that has a current carrying capacity of at least 20 A and this will be 2.5 mm^2 cable. This is found using vertical column 6.

Once we have established the size of the cable we are to use we must also refer to the continuation of 4D2B which gives the voltage drop for cables in millivolts per ampere per metre. For a 2.5 mm^2 cable we get a value of 18 mV/A/m from column 3. We can now use this value along with the other information we have to determine the voltage drop along the length of the cable.

To find the volt drop we need to know the length of the cable, the amount of current it will carry and the volt drop per ampere per metre.
We put this into the formula as

$$\text{Volt drop} = \frac{\text{mV}/\text{A}/\text{m} \times I_b \times \text{length}}{1000}$$

This will give us a volt drop in volts.
In our example

$$\text{Volt drop} = \frac{18 \times 16 \times 20}{1000} = 5.76$$

Remember: the load current, I_b, is used to calculate voltage drop.

So now we know the actual voltage dropped along the length of the cable but does this comply with the maximum allowed? To find this out we must determine the maximum value. We will assume that for our calculation the voltage drop measured from the main intake position to any point on the installation must not exceed 4% of the nominal supply voltage. This is the maximum allowed in BS 7671:1992.

For our installation this is 230 V and so we have

$$\frac{4 \times 230}{100} = 9.2 \text{ V}$$

The actual volt drop must be equal to or less than the maximum volt drop allowed and in our example 5.76 V< 9.2 V so our circuit does comply with the volt drop constraints.

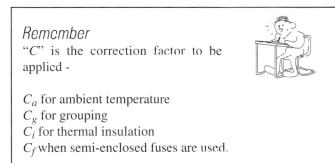

Figure 5.3   *If the maximum volt drop is 9.2 V then the volt drop from "C" to "D" must not be greater than 9.2 − (1.5 + 1.0) − 6.7 V*

Now let's have a go at a more complex installation.

A circuit is to supply a load of 2.75 kW with a power factor of 0.9 at 230 V 50 Hz. A PVC insulated and sheathed steel wire armoured cable is to be installed clipped to a perforated metal cable tray with a total cable length of 30m from the distribution board. The distribution board is located at the supply intake position. For most of its length the cable is installed touching two other cables on an ambient temperature of 40 °C. If protection is provided by a BSEN 60269-1 (BS88) type fuse what is the minimum size of cable that can be used to comply with both current carrying capacity and voltage drop constraint if the equipment will not function at less than 224 V?

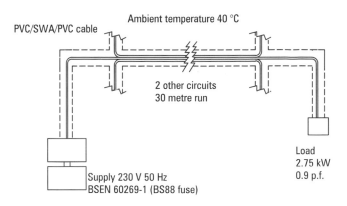

Figure 5.4

Solution: The reference method from Table 4A is method 11.

Power	=	$U \times I_b \times$ power factor
2750	=	$230 \times I_b \times 0.9$
I_b	=	$\dfrac{2750}{(230 \times 0.9)}$
	=	$\dfrac{2750}{207}$
	=	13.285 A

$I_b = 13.285A \quad \therefore I_n \geq 13.285$

from Table 41B1, for type BSEN 60269-1 (BS88) fuses, $I_n = $ 16 A

$$I_b = \frac{I_n}{C_a \times C_g \times C_f \times C_i}$$

C_a	=	0.87 from Table 4C1
C_g	=	0.81 from Table 4B1
C_f	=	1.0 as a BS3036 type is **not** being used.
C_i	=	1.0 as cable is not surrounded by insulation

I_t	=	$\dfrac{16}{0.87 \times 0.81 \times 1 \times 1}$
	=	$\dfrac{16}{0.7047}$
	=	22.7A

Select cable size from 4D4A column 4.

1.5 mm^2 at 22 A is too small so we must use 2.5 mm^2 at a rating of 31 A.

$$I_z = 31 \text{ A}$$

mV/A/m from Table 4D4B is 18 mV

Maximum volt drop = 230 – 224 = 6 V

Actual volt drop

$$= \text{mV/A/m} \times I_b \times \text{length}$$

$$= \frac{18 \times 13.285 \times 30}{1000}$$

$$= 7.17 \text{ V}$$

This does not comply as 7.17 > 6 V (the maximum allowed).

We must now go up in cable size in order to reduce the volt drop.

4 mm^2 has a mV/A/m of 11 mV

$$\therefore \text{volt drop} = \frac{11 \times 13.285 \times 30}{1000}$$

$$= 4.384 \text{ V}$$

As 4.384 V < 6 V maximum then this cable will be suitable.

So the smallest cable that we can use to ensure compliance with both current carrying capacity and voltage drop constraints is a 4 mm^2.

In this example we had to carry out the same calculation for voltage drop twice as the first selection did not comply. In some instances we may find that we do this calculation 3 or more times. We can avoid this by taking an alternative approach.

Once we know the values of I_b, length and the maximum permissible volt drop we can calculate the maximum value of mV/A/m that will comply with the volt drop requirements.

By using the formula

Maximum volt drop

$$= \frac{\text{mV/A/m} \times I_b \times \text{length}}{1000}$$

we can see that

Maximum mV/A/m

$$= \frac{\text{maximum volt drop} \times 1000}{(I_b \times \text{length})}$$

If we use our example

Maximum mV/A/m

$$= \frac{6 \times 1000}{13.285 \times 30}$$

$$= 15.05 \text{ mV/A/m}$$

Now when we select our cable size and go to the mV/A/m table we can see that as the maximum acceptable value would be 15.05 mV and 2.5 mm^2 has a value of 18 mV that it would not be suitable. We then continue down the column until we come to a value that is equal to or less than our maximum. By using this method we only have to carry out the calculation for volt drop once and then make a selection from the tables.

Try this

A circuit is to supply a load of 6 kW and unity power factor at 230 V 50 Hz a distance of 10 m from the main intake position. As a degree of mechanical protection is required a PVC insulated steel wire armoured PVC served cable is to be used and it is to be protected by a BSEN 60898 Type B MCB. The cable has to run through an area with an ambient temperature of 35 °C and is installed touching one other cable throughout its length. If the cable is to be clipped direct to a perforated steel cable tray determine the minimum size of the cable that can be used to comply with current carrying capacity and voltage drop constraints.

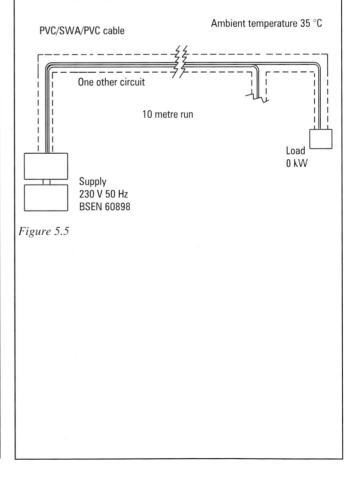

Figure 5.5

We have found that we may need to increase the cross sectional area of cables to comply with the volt drop constraint. In practice cable sizes are often dictated by volt drop rather than the current carrying capacity. It is therefore vital that this check is made and that selection is not done on current carrying capacity alone.

Points to remember

Having calculated the current carrying capacity required and selected the conductor size it is vital that the_____ _____ calculations is carried out. This is achieved using the mV/A/m value from BS 7671:1992 and the I_b value. Cable sizes are often determined by v_____ d_____constraint rather than c_____ c_____ c_____.

The actual volt drop must be _____ ____ or ____ _____ _____ the maximum volt drop allowed.

A cable touching three other cables has a grouping factor for _____ cables touching and this factor can be found in _____.

It is not uncommon to supply a 10 A load with a 2.5 mm² cable if the run is of considerable length. A maximum voltage drop of 4% of the nominal supply voltage is permitted from the supply intake position.

PROJECT

Refer to the specification and drawings in the Appendix.

This example calculation refers to the cooker circuit in the project (Circuit 1).

Installation method
  Installation method 3 from Table 4A Item 3

Select cable
  $I_z \geq I_t$

  ∴ from Table 4D1A I_z = 57 A
  from Col 4 for a 10mm² conductor with
  mV/A/m of 4.4 mV

Determine actual voltage drop
  Design current   = 35.416 A
  Length of run   = 17.3 m
  Voltage drop   = mV/A/m × I_b × l
                    $= \dfrac{4.4 \times 35.416 \times 17.3}{1000}$
                    = 2.6958 volts
  Voltage drop   = 2.7 volts

  Maximum voltage drop   = 4% of nominal supply voltage
  ∴ Maximum voltage drop = 4 % × 230
                    = 9.2 V
  ∴ Circuit complies with voltage drop constraints.

Now, using Circuit 2, determine the installation method, select the cable, determine the actual voltage drop and the maximum voltage drop assuming 4% of the supply and state whether the circuit complies with voltage drop constraints.

Project – Circuit 2

1. State four factors that will affect the volt drop over a length of cable.

2. If a cable carries a current of 15 A over a distance of 25 m and the maximum voltage drop allowed is 3 V calculate the maximum mV/A/m that would give compliance with the volt drop constraints.

3. A circuit supplies a load of 4 kW at a power factor of 0.95 over a distance of 25 m measured from a distribution board located at the main intake position. The supply is 230 V, 50 Hz and the circuit is protected by a BS1361 fuse. Determine the minimum size of light gauge MICC/PVC sheathed cable required to comply with current carrying and a maximum voltage drop of 6 V. The cable is installed clipped directly to a non metallic surface, exposed to the touch, and there are no factors to be applied.

4. A PVC insulated, LSF sheathed SWA cable has two conductors of 16 mm^2 copper. It is installed from the main incoming supply position to an isolator 50m away clipped to a perforated steel cable tray. For most of its length it is run spaced from two other cables in an ambient temperature of 25 °C. The circuit is protected by a BSEN 60269-1 (BS88) type fuse. What is the maximum load that could be connected to the isolator to ensure that the voltage drop due to the cable does not exceed 3 V?

6

Selection of Protective Conductors

Check that you can remember the following facts from the previous chapter.

____ is the tabulated current carrying capacity of the cable

____ is the tabulated current found in the tables

____ is the design current of the circuit

____ is the operating current of the fuse or circuit breaker

Put the appropriate sign ($\geq$ or $\leq$) in the following:

$$I_z \rule{2cm}{0.4pt} I_t$$

$$I_n \rule{2cm}{0.4pt} I_t$$

$$I_b \rule{2cm}{0.4pt} I_n$$

The maximum volt drop allowed in BS7671 must not exceed 4% of what?

On completion of this chapter you should be able to:

- define the terms "exposed conductive part" and "extraneous conductive part"
- state the requirements for shock protection for circuits supplying both socket outlets and fixed equipment
- calculate the prospective earth fault current from given data
- establish the disconnection time of protective devices under earth fault conditions
- check circuits for compliance with the requirements for shock protection
- state what is meant by thermal constraint and identify the criteria necessary to calculate compliance
- carry out calculations to find the minimum size of circuit protective conductor

Part 1

In the previous chapter we established the requirements for selecting the size of cables based on the live conductors. We shall now determine the method of sizing the protective conductors for our circuits to ensure compliance with the requirements for shock protection. In order for us to understand the requirements for shock protection we must be familiar with the effects of an earth fault on

- the equipment supplied
- the whole of the installation
- the building structure

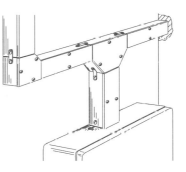

Figure 6.1   All cpcs must be continuous throughout.

Before we can begin there are a couple of terms that we must be familiar with. These define two components that may become part of the earth fault path and they are "Exposed Conductive Part" and "Extraneous Conductive Part". The definition of each of these terms can be found in BS 7671:1992.

Exposed conductive part

An exposed conductive part is defined as:

"A conductive part of equipment which can be touched and which is not a live part but which may become live under fault conditions".

This means any exposed metal parts of the electrical installation. This is because any of these parts can become live in the event of an earth fault on any circuit within the installation as all the circuit protective conductors are connected to a common point. If the potential on any of these conductors rises above earth potential then all the parts connected to the common point will also rise above earth potential.

Figure 6.2

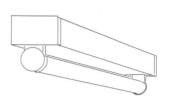

Figure 6.3   *The metal case of a motor or fluorescent luminaire is an exposed conductive part.*

Extraneous conductive part

The definition of an extraneous conductive part is given as

"A conductive part liable to introduce a potential, generally earth potential, and **NOT** forming part of the electrical installation".

This means any structural steelwork, water pipes, gas pipes, drain pipes - in fact it can include any metalwork within the confines of the installation that is not a part of the electrical installation. We must satisfy ourselves as to the probability of the metalwork introducing a potential, including earth potential, to any point within the installation, this will establish whether there is a need to bond these parts to the common earth point.

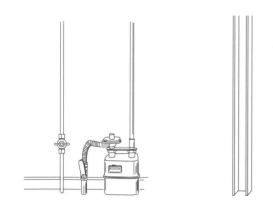

Figure 6.4   *Metal services and structural steel are extraneous conductive parts.*

Gas and water pipes will need to be protected against the effects of an earth fault current. Structural steel may or may not need this protection.

In order to protect both the installation and the consumer against the effects of an earth fault we must provide a safe return path for any earth fault currents. There are a number of ways to provide this return path and the method used will depend on the type of supply system to which the installation is connected. When a fault to earth occurs current flows around the earth fault loop (the safe return that we provide for faults). At this point it would be as well to remember that we actually require the highest possible current to flow. This will ensure the rapid operation of the protective device and the rapid disconnection of the circuit from the supply.

Earth fault path

The impedance of the earth fault path, known as Z_s, plays an important part in the system as it will regulate the amount of current that flows in the earth fault path.

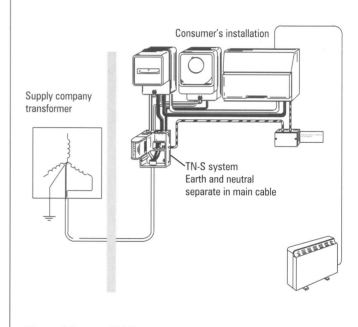

Figure 6.5   *TN-S system*
Earth faults are returned to the supply transformer through the metal sheath of the supply cables

If we look at the circuit diagram in Figure 6.6 we can see the case of the electric heater is connected to the circuit protective conductor. This is in turn connected to the consumer's earth terminal and then via the earthing conductor and the supply system to the star point of the transformer. All of which is best shown on the TN-S system for clarity.

All the parts of the circuit which are the consumer's responsibility, and connected to the consumer's earth terminal, are "exposed conductive parts". These are all part of the electrical installation and include conduit, trunking, circuit protective conductors and cases of appliances and equipment. Remember that in the case of a fault to earth all the "exposed conductive parts" of the installation become live for the period of time that it takes for the protective device to disconnect the circuit from the supply. The earth fault current will flow around the earth fault loop as shown in Figure 6.6.

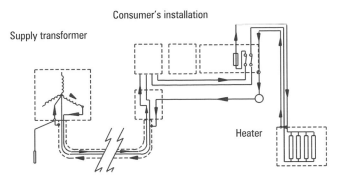

Figure 6.6   *Circuit diagram for TN-S system*
Arrows show the path that the current will take
in the event of a fault.

This path will offer some resistance to the flow of current which will be dependent on the impedance of the conductors which go to make up the loop. As we can see, the loop comprises the transformer winding, the phase conductor of the supply and the consumer's circuit up to the fault. From the fault to the consumer's earth terminal is the circuit protective conductor and from the consumer's earth terminal, via the earthing conductor, back to the star point is dependent on the type of system. For our calculations we assume that the fault itself offers no resistance to the flow of current.

Figures 6.7 and 6.8 show the two other main types of system used in the public supply system in the UK.

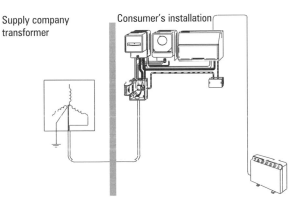

Figure 6.7   *TN-C-S system*
The supply cable has a combined earth and
neutral conductor.

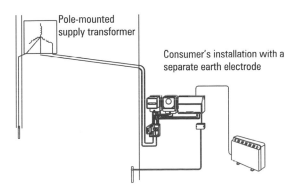

Figure 6.8   *Where the supply company is unable to provide*
an earth return to the transformer, the consumer
requires a separate earth electrode.
An RCD is the preferred type of protective
device to ensure adequate protection.

In order for us to determine the current that will flow in the event of a fault we need to establish the impedance of this earth fault loop. If the installation is already installed then the earth fault loop can be measured using a line-earth loop impedance tester. If we are designing a system then we need to calculate the value of earth fault loop.

To do this we use the value of earth fault loop impedance of the supply system which we obtained from the supply authority, known as Z_e. To this we must add the impedance of the part of the loop that is made up by the circuit conductors. This will be the phase conductor up to the fault and the circuit protective conductor back to the consumer's earth terminal. These values are known as R_1 and R_2 respectively and are taken to the point on the circuit furthest from the supply to establish the worst case, that is, when the conductor impedance is at its maximum.

We can best see how this is done by using our example of the electric heater and calculating the earth fault loop impedance.

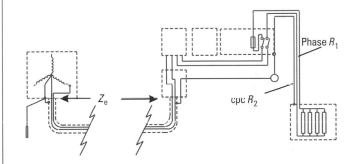

Figure 6.9

Conductor resistances

From earlier calculations we can assume we have established that the size of the live conductors we are using as 2.5 mm². The standard sizes of composite cables are supplied with the cpc one size smaller than the live conductors, so it would be a good idea to take this as a starting point for determining the value of the earth fault loop impedance. We know then that the size of the cpc is going to be 1.5 mm² and so we are ready to calculate the value of the earth fault loop impedance of the consumer's part of the system (R_1 and R_2).

To carry out the calculation we need to know the following details

- the length of the circuit conductors
- the cross sectional area of the phase conductors
- the cross sectional area of the circuit protective conductor (this may be the same as that of the phase conductor but not always)
- the impedance of the phase and protective conductor per metre

This last detail we can get from tables, so let's take a look at the section of these as shown in Table G1 in Guidance Note 1.

Try this

Complete the tables below using tables in IEE Guidance Note 1.

Table 6.1 *Values of resistance/metre for copper conductors at 20 °C in milliohms/metre.*

Cross-sectional area mm²		Resistance milliohms/metre
Phase conductor	Protective conductor	Plain copper
1	–	
1.5	–	
1.5	1.5	
2.5	–	
2.5	1.5	
2.5	2.5	
4	–	
4	2.5	
6	2.5	
6	4	

To allow for temperature rise under fault conditions the following correction factors must be applied.

Table 6.2

Insulation material	70 °C PVC
Multiplier	

The important thing to remember is that the values given are in MILLIOHMS per metre.

For our calculation we require the resistance of a 2.5 mm² phase conductor with a 1.5 mm² protective conductor. From Table 6.1 this is 19.51 mΩ/m. This is not the end of the calculation though as we need to know the resistance under fault conditions and this could mean the temperature rising in the conductors and increasing their resistance. To allow for this a factor of 1.2 must be applied to our total value.

We now have all the details that we need to calculate the value of the earth fault path within the consumer's installation. This will be $R_1 + R_2$.

Assume the length of run = 20 m
resistance from Table 6.1 = 19.51 mΩ/m
multiplier from Table 6.2 = 1.2

Remember: Always use the multiplier from Table 6.2.

The total value will be calculated using the formula:

$$\frac{\text{resistance}}{(R_1 + R_2)} = \frac{m\Omega / m \times \text{length} \times \text{multiplier}}{1000}$$

$$\text{resistance} = \frac{19.51 \times 20 \times 1.2}{1000}$$

$$= 0.46824 \ \Omega$$

Try this

Determine the value of $R_1 + R_2$ from Table 6.1 for

1. 1.5 mm² phase conductor with 1.5 mm² cpc

2. 2.5 mm² phase conductor with 2.5 mm² cpc

3. 6.00 mm² phase conductor with 2.5 mm² cpc

Earth fault loop impedance value

To establish the total value of earth fault loop impedance we must add to this the earth fault loop impedance of the supply, Z_e. We shall assume the supply company has quoted a value of 0.35 Ω for this, so we get a total of

$$0.35 + 0.46824 = 0.81824 \ \Omega$$

The earth fault loop impedance is found using the formula

$$Z_s = Z_e + (R_1 + R_2)$$

So what is the significance of the value of Z_s?

If we refer to tables 41B1, 41B2 and 41D in BS 7671:1992 we find the types and ratings of protective devices listed. Below each rating is given a maximum value for Z_s for the device.

Tables 41B1 and 41B2 give the maximum values of Z_s for devices supplying circuits incorporating socket outlets. Tables 41D and 41B2 give maximum values for circuits supplying fixed equipment.

We can see that the maximum values of Z_s are different for each table. This is because the disconnection time required for a circuit supplying socket outlets is given as within 0.4 seconds. Circuits supplying fixed equipment must disconnect within 5.0 seconds. The value of Z_s for fixed equipment can therefore be higher than that for socket outlets as a longer period for disconnection is allowed, therefore a lower current can flow.

We will assume that on our circuit supplying the electric heater we are using a 15 A BS1361 type fuse to supply fixed equipment. We use Table 41D to check for compliance of the circuit. The maximum value of Z_s is given in the table as 5.22 Ω for a 15 A BS1361 type fuse. As our value is 0.81824 Ω, this circuit does comply.

If we apply a simple Ohm's law calculation to our circuit then we can find the current that will flow in the event of a fault to earth. This is known as the prospective earth fault current I_f and is found by using the formula

$$I_f = \frac{U_o}{Z_s}$$

where:

I_f is the prospective fault current

U_o is the nominal phase voltage to earth and

Z_s is the earth fault loop impedance

so $\quad I_f \quad = \quad \dfrac{230}{0.81824} \quad = 281\ A$

Now that we know the value of the prospective fault current, we can further check the compliance of the circuit by using the tables in Appendix 3 of BS 7671:1992. These tables are the simplified time/current curves for the various types of protective devices.

The horizontal axis is the prospective fault current and the vertical axis is the time. The divisions along the axis are logarithmic and, as we can see, the divisions are not equal in size. Figure 6.10 shows a section of the X (horizontal) axis of the graph. We can see that the origin of the axis, the extreme left-hand end, has a value of 1 NOT 0 and the divisions go up in units of one until we reach a value of 10. The divisions then go up in units of 10 until a value of 100 is reached at which point the divisions go up in units of 100 and so on. In this way a considerable current range can be covered in a relatively small space.

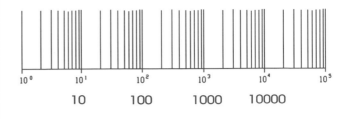

Figure 6.10

Figure 6.11 shows a section of the X axis where we've had to subdivide one of the sections. We must remember that subdivisions of scale sections must be done in the same ratio as the main scale. This means that the value 250 A is NOT going to be halfway between 200 A and 300 A. It occurs at a point approximately 0.66 of the way between these two as shown on Figure 6.11.

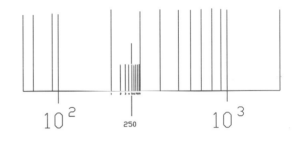

Figure 6.11   *The value 250 is not halfway between 200 and 300 but nearer two-thirds.*

The vertical Y axis is shown in Figure 6.12 and again we can see that the origin of the axis is NOT 0. For ease of location the 1 second position is highlighted on the axis in Figure 6.12. By working back down the axis we can see that the origin of the axis is in fact 0.01 seconds. Like the X axis the divisions are logarithmic and so the same proportional division of the scale must be used.

Try this

Mark the positions on the scale of the following currents: 6 A, 550 A, 65 A, 1250 A, 275 A

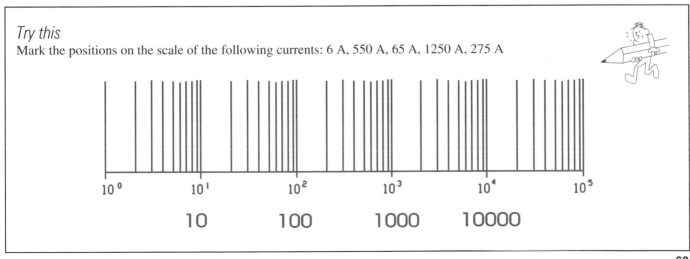

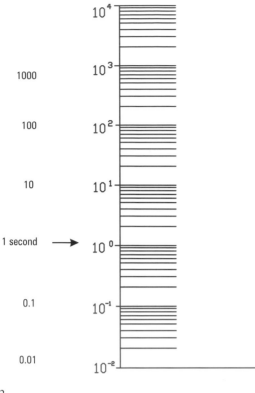

Figure 6.12

Now to further check the compliance of our circuit we can use the value of I_f and the appropriate time/current curve to establish the disconnection time for the device. To do this we use the appropriate table for BS1361 fuses in Appendix 3, BS 7671 or a manufacturer's details. Figure 6.13 shows the appropriate section of the graph.

We move along the X axis until we reach the value of I_f for our circuit (281 A), we then move vertically up from this point until we bisect the curve for the 15A device. A rule will be of some help when carrying out this exercise. We then move horizontally across from this point to the Y axis and read off the time. As the circuit we are considering supplies fixed equipment the disconnection time must be no more than 5 seconds. We can see that a 15 A BS1361 type fuse requires a current flow of approximately 67 A to achieve disconnection within 0.1 sec. The value we get with a current of 281 A is in fact less than 0.1 seconds so our circuit complies with requirements, but we expected this as the maximum value of Z_s already indicated this to be the case.

TIP

Use a highlighter to mark across the curves at the 5 s and 0.4 s positions for ease of reference.

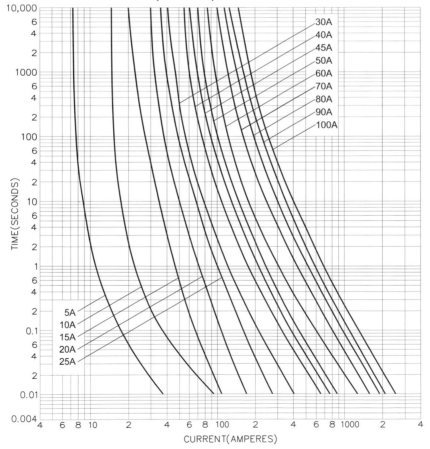

Figure 6.13   Time/current characteristics for fuses to BS 1361.
Reproduced with kind permission of Bussmann Division, Cooper (U.K.) Ltd.

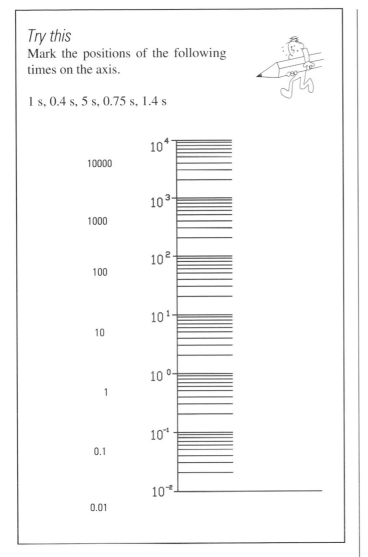
Part 2

Up to now we have found the size of conductors needed to give compliance with requirements for shock protection. Whilst doing this we found that a high current flows in the event of an earth fault under the correct conditions. To provide shock protection this current must be carried by the circuit protective conductor for the time that it takes for the protective device to operate and disconnect the circuit from the supply. As the circuit protective conductor is generally a smaller cross sectional area than the live conductors, and the current that it carries may be quite high, a great deal of heat will be generated whilst the current flows.

We must make sure that, whilst the fault current is flowing, the heat produced will not cause damage to the conductors or the insulation and material surrounding them. If sufficient current flows for long enough the heat produced could be such that the insulation catches fire. Once this happens disconnection of the supply will not extinguish the flames.

Obviously this situation cannot be allowed to occur. To prevent it we must ensure that the circuit protective conductor is large enough to carry the fault current for the period of time needed for the device to disconnect the supply without excessive heat being produced.

The thermal constraint placed on the circuit by BS 7671 ensures that the circuit protective conductor of the circuit is large enough to carry the earth fault current without a detrimental effect on the conductor, the insulation or the installation.

In order to find the minimum cross sectional area of the circuit protective conductor we require the following information
- the prospective earth fault current "I_f"
- the time taken for the protective device to operate with this fault current "t"
- the constant k which is related to the type of circuit protective conductor and its method of installation

The way in which these are related to the minimum size of the circuit protective conductor is by a formula known as the adiabatic equation.

The minimum cross sectional area of the conductor is known as "S" and the formula we use is

$$S = \frac{\sqrt{I^2 \times t}}{k}$$

(The value of I_f is used for this calculation.)

It is important that the calculation is carried out in the right order to give the correct answer.

Let's consider the circuit supplying the electric heater to see if it complies with the requirements for thermal constraints. We know that the prospective earth fault current is 281 A. We also found that disconnection time will be less than 0.01 of a second. Now we need to find the value of the constant k for our circuit. The values for k are given in tables 54B to 54F in BS 7671:1992. At the head of each table is the description of the method of installation of the circuit protective conductor. For our type of cable and the method of installation we find the value of k from Table 54C to be 115.

If we now put these values into the formula we have

$$S = \frac{\sqrt{281^2 \times 0.1}}{115}$$

We must carry out the calculation in the correct sequence to give the correct solution

stage one: Square the value of I_f, i.e. $I_f \times I_f$

stage two: Multiply the result by the time
               t second

stage three: Take the square root of the result

stage four: divide the answer by the value of k

In our particular case this would be

281×281 $= 78961$

7896×0.1 $= 7896.1$

$\sqrt{7896.1}$ $= 88.86$

$\dfrac{88.86}{115}$ $= 0.773 \text{ mm}^2$

As the minimum size of cpc to give compliance is 0.773 mm^2 and we have installed a cpc of 1.5 mm^2 we can see that circuit fulfils the requirements for thermal constraints and so our circuit complies.

If the size of cpc installed proves to be insufficient then a larger cross-sectional area conductor must be used. This will, of course, have an effect on the value of the earth fault loop impedance as a larger cpc will reduce the impedance and a higher current will flow. This in turn will reduce the disconnection time.

We could calculate the minimum requirements based on the known criteria that apply to our circuit. For example we know that for a socket outlet circuit the maximum disconnection time is 0.4 seconds. If we know the type and rating of the protective device we can establish the minimum value of I_f to give disconnection in 0.4 seconds. This data and the value of k for the type of cpc and its method of installation will allow us to establish the minimum size of cpc to give compliance under the worst conditions.

To do this we must use the minimum value of I_f to give the disconnection time required, the maximum disconnection time and k for the type and method of cpc installation. If we carry out this one calculation we can establish the minimum size of cpc to comply with the absolute worst conditions and we can ensure that the size selected will be within a usable range. This can be particularly useful when designing circuits for installation in conduit and trunking where the size of cpc can be varied with ease. It will also ensure that a great deal of time is not wasted carrying out repeated calculations to establish an acceptable size.

Points to remember ◄ $- - - - - - - - - - - - -$

An exposed conductive part of equipment which can be touched and which may become live under _____ conditions.

Two examples of exposed conductive parts are:

An extraneous conductive part is a conductive part liable to introduce a _____, generally earth _____, and NOT forming part of the electrical installation.

Two examples of extraneous conductive parts are:

The earth fault path is a safe return path provided for any _____ _____ _____.

In the event of an earth fault we require the highest possible current to flow. Explain why.

What are the four factors we need to know before calculating the value of the earth fault loop impedance?

The earth fault loop impedance Z_s determines the current flow to earth in the event of a _____.

Z_s is made up of $Z_e + (R_1 + R_2)$

Disconnection time is dependent upon fault current and type of _____ _____.

The values of I_f, t and the constant k are used to determine the minimum size of cpc. We use the formula

$$S = \frac{\sqrt{I^2 \times t}}{k}$$

where S = _____ _____ _____ _____

The example calculation refers to the cooker circuit (Circuit 1) in the project.

Refer to the specification and drawings shown in the Appendix.

Maximum value of earth fault loop impedance Z_s (from Table 41D) is 2.4 Ω.

Actual earth fault loop impedance
$$\text{actual value of } Z_s = Z_e + (R_1 + R_2)$$

Remember for steel conduit as a conductor the maximum value of impedance to comply with the British Standards is 0.005 Ω per metre.

So

$$
\begin{aligned}
Z_e &= 0.3\ \Omega \\
R_1 &= 1.83\ \text{m}\Omega \text{ from Table 6.1, multiplier is 1.20} \\
&\quad \text{from Table 6.2} \times 17.3\ \text{m} \\
R_2 &= 0.005\ \Omega \text{ (worst case)} \times 17.3\ \text{m}
\end{aligned}
$$

$$\therefore Z_s = 0.3 + \underbrace{\frac{(1.83 \times 1.20 \times 17.3)}{1000}}_{R_1} + \underbrace{(0.005 \times 17.3)}_{R_2}$$

$$
\begin{aligned}
&= 0.3 + 0.04244908 + 0.0865 \\
Z_s &= 0.4244908\ \Omega \qquad \approx 0.4245\ \Omega
\end{aligned}
$$

$\therefore$ Circuit will comply with shock protection requirements.

Earth fault current I_f

$$I_f = \frac{U_o}{Z_s} = \frac{230}{0.4245\ \Omega}$$

$$= 541.81\ \text{A}$$

Disconnection time from Fig 3B in Appendix 3

$$t < 0.1\ \text{s}$$

Minimum size of cpc

$$S = \frac{\sqrt{I^2 \times t}}{k}$$

Where S = minimum cross sectional area of cpc
I = earth fault current I_f
t = disconnection time
k = constant from Table 54E

$$S = \frac{\sqrt{541.81^2 \times 0.1}}{47}$$

$$S = \frac{\sqrt{29355.81}}{47}$$

$$S = \frac{171.34}{47}$$

$$S = 3.65\ \text{mm}^2$$

Actual cross sectional area of cpc

Outside diameter of 20 mm conduit = 20 mm
Inside diameter of 20 mm conduit = 17 mm
$\therefore$ c.s.a. of 20 mm diameter –
c.s.a. of 17 mm diameter = c.s.a. of cpc

$$\frac{\pi \times 20^2}{4} - \frac{\pi \times 17^2}{4} = 314.159 - 226.98 = 87.179$$

Cross sectional area of steel = 87.18 mm^2

As steel is less conductive than copper an allowance must be made for this in the c.s.a. It is a good rule of thumb to consider the ratio as 10:1 where the steel conductor requires a c.s.a. 10 times larger than the copper. In this example then our steel is a copper equivalent of around 8.7 mm^2.

Now carry out the final calculations for Circuit 2 to determine the protective conductor.

Self-assessment short answer questions

1. A 230 V circuit is to supply a domestic electric cooker with a socket outlet fitted to the control panel with a load of 27 A after diversity has been applied. The cable is to be a PVC/PVC composite cable with a 6 mm^2 phase conductor and a 4 mm^2 cpc to a location 25 m from the supply intake position and no factors need to be applied. It is protected by a BSEN 60898 type C circuit breaker and Z_e is 0.4 Ω. Determine the minimum size of cpc that could be installed to comply with the requirements for thermal constraints.

2. A circuit supplies an item of fixed equipment from a 230 V 50 Hz supply. If the circuit protective conductor is 2.5mm^2, the value of k is 115 and the disconnection time is to be 0.4 seconds what is the maximum value of earth fault current that can flow for compliance with the thermal constraint requirement?

3. Determine the minimum cross sectional area of protective conductor for a 230 V single-phase circuit which has the following?
   i)   a value for Z_e of 0.3 Ω
   ii) a value for $R_1 + R_2$ of 0.7 Ω
   iii) a circuit protective device of 30 A to BS1361
   iv) a k factor of 143

4. a. What is the minimum fault current that needs to flow through a 100 A fuse to BS 88 parts 2 & 6 if it is to operate in 0.4 seconds?
   b. If the protection device in (a) was replaced with a fuse to BS 3036 when the same fault current was flowing how long would it take the new fuse to operate?

7

Preparing for Commissioning

Check that you can remember the following facts from the previous chapter.

The earth fault path consists of the circuit protective conductor and the earth resistance external to the installation. To complete the circuit the phase conductor is also included in the resistance path.

As the circuit external to the installation is in the hands of the _____ _____ this is all put together as one value covering the impedance of the earth circuit, transformer winding and phase conductor. This impedance is represented by _____, the external impedance. The formula to calculate the cross-sectional area of the circuit protective conductor is

where

$S =$

$I =$

$t =$

$k =$

This formula takes into account the fact that heat is produced when a fault current flows and this heat can damage the insulation of the cable if it is not cleared within the appropriate time.

On completion of this chapter you should be able to:

◆ recognise the need for a good quality of workmanship
◆ recognise why inspections and tests should be carried out
◆ identify when to carry out inspection and tests
◆ recognise the need to carry out inspections on
   – cables and switchgear
   – switchgear and accessories
   – earthing and bonding
◆ relate the inspection to the type of circuit

Part 1

Quality

Whenever we purchase electrical equipment, we expect it to be designed, made and tested to the appropriate standards. These standards show that the equipment at least complies with minimum safety requirements.

There are also standards of quality. Many manufacturers now state that their products are manufactured in accordance with BS5750. This is the UK national standard for quality systems, governing every single aspect of the manufacturing process.

Quality should not stop at electrical equipment. The aim should be to produce a good quality electrical system. This includes the work of those involved in design, installation and testing.

Monitoring

To ensure a level of quality, standards and acceptable levels have to be identified. These apply to all aspects including - the design, equipment and workmanship. If any one of these is not to the acceptable standard then the overall quality is lower than it should be. To monitor quality a system needs to include methods of measuring the acceptable standards. In electrical installations these methods are usually a mixture of - statutory requirements, recommendations from learned bodies and the customer's requirements.

Namely:
• The Electricity Supply Regulations 1988
• The Electricity at Work Regulations 1989
• BS 7671
• Relevant British Standards
• The customer's specifications
• relevant European Standards
Identifying the standards is one thing, ensuring they are being adhered to is another. One way of monitoring standards is by using check lists and recording when and where the checks are carried out.

In electrical installations there are many stages that have to be gone through before the customer can use the installation. An electrical installation is the result of work by several different experts. On small installations this may only involve one or two people but on large ones this could involve hundreds.

When the electrical system is being installed, the electricians are working to drawings and specifications produced by architects and electrical designers.

These experts have to take many factors into account to produce the necessary design documentation. In addition to statutory and non statutory regulations there are many requirements of equipment manufacturers. Safety of both the installer and future user must always be given a high priority. As the design can be extremely complex, the installation staff must have sufficient knowledge and supervision to carry out the work to the design specification. If the installation is not carried out to the original design there can be no guarantee as to its safety or correct working.

To ensure the installation is safe to use and operates as the customer intended, a schedule of acceptable standards must be drawn up. When this has been done a schedule of everything that has to be tested must be listed. The acceptable standards are usually taken from BS 7671, with changes to meet the customer's requirements. Schedules of the work involved in the inspection and testing can be compiled on forms such as those shown in BS 7671 and IEE Guidance Note 3. When the schedules have been compiled, the record sheets can be prepared with the acceptable standards being filled in. It is not until all of this has been completed that tests can be carried out to see if the installation meets the acceptable standards.

This process does not only apply to new installations but also to old installations and whenever alterations or additions are made to an existing installation.

Remember
Inspection and testing by themselves do not ensure quality. They are only a part of the process.

New installations
At the completion of the installation a thorough inspection of the work needs to be carried out and then tests made to ensure that all of the necessary standards have been upheld. It is only after this has been carried out and any defects have been corrected that the installation can be said to be complete and safe to use.

Old installations
After the installation has been in use for some time, further inspections should be carried out to ensure the standard of the installation has not deteriorated. This should then be followed by tests and the results of these compared with the original tests and the design. If results vary then further investigation needs to be carried out to find out why.

Alterations
Changes are often made during the life of an installation. These may be changes that involve moving existing parts of the installation or adding new sections on. In either case care has to be taken to ensure that the original design has not been changed in such a way as to make it dangerous. To confirm this, the new work and any associated wiring must be inspected, tested and compared to the original and any new designs to check that it complies.

Where the original design has been amended, the test results should be checked against this. If, however, very little thought has been given to the original design and the alterations may have changed it, the tests must then go further. This is to ensure that original design factors have not been changed in such a way that hazards have been introduced. Often in these cases the test engineer has to work back and prove the original design met acceptable standards.

When to inspect and test
So inspections and confirming tests have to be carried out at different times depending on the circumstances. These times can be listed as
- continuous as work is carried out
- on completion of an installation
- at recommended intervals after completion
- when any additions or alterations are made

Why inspect and test?
The inspections and tests are carried out to confirm that the installation
- meets the requirements laid down in the design
- meets all legal requirements and relevant non statutory regulations
- equipment is to the appropriate British Standards
- is not damaged in any way so as to be dangerous
- meets the requirements of the equipment manufacturer

Try this
Explain, in your own words, the main purposes for carrying out inspections and tests on any electrical installation.

Inspection

This is often referred to as the visual inspection as in most cases it involves **LOOKING** at parts of the installation. It should not however preclude the use of the senses of smell and hearing. When some electrical equipment is overloaded and gets hot it gives off a very distinctive smell. This can lead to possible dangers before any visible signs are apparent. Similarly when there are loose connections these often arc and crackle. These again can lead to possible fire hazards before there are any visible signs. Touch can identify equipment running warm, the presence of moisture or corrosion. So although most of an inspection is visual other senses should also be alert to possible dangers.

An inspection should be carried out at different stages in the life of an installation and it will cover the different aspects of the installation such as:
- the cables
- the cable enclosures, such as conduit or trunking
- all switchgear and control gear
- accessories
- earthing and bonding arrangements.

Cables

Cables need to be inspected to see if they have been properly installed. We need to check for adequate suitable fixings and to ensure that the cable supports are in place. Supports systems and cleats may become loose or damaged by, for example, mechanical impact or corrosion. Terminations, both of the cable and the conductors, should be mechanically and electrically sound. Where cables can be seen, they should be visually inspected for damage from heat, corrosion or mechanical impact and obvious signs of deterioration.

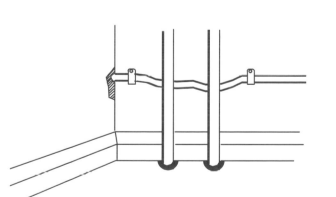

Figure 7.1   Heating pipes have been added which may have not only damaged the cable but may have made it necessary to re-site it.

Try this

Taking the cover off a ceiling rose can expose a multitude of "sins"!

List the "sins" that you can identify in the figure below.

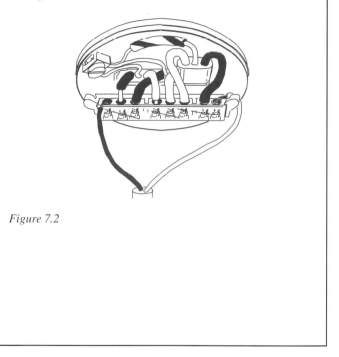

Figure 7.2

Cable enclosures

Conduit installations need to be checked for different things depending on whether it is steel or PVC conduit. If it is steel conduit then corrosion can be a problem so the conduit should be examined to see if the finish of it is suitable for the environment. When cutting or tightening threads steel conduit can get scour marks which should be cleaned off and painted. A visual inspection may need to be carried out to ensure this has been done.

Figure 7.3   Conduit, if it does not have a suitable finish for the environment, may corrode.

The fixing of conduit can be a problem because of its weight and what accessories are connected to it. An inspection which looks at how well the conduit is fixed to the surface should also consider the possible temperature changes which may result in the fixings changing their characteristics. PVC conduit is particularly prone to temperature change and expansion joints should be fitted to allow for this. As the conduit is a method of mechanical protection, it should be complete with all covers in place and any unused holes blanked off.

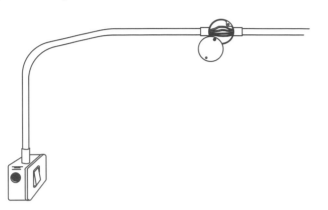

Figure 7.4   *All holes not in use should be plugged and box lids and the like should be securely fixed in place.*

Steel conduit, whether it is used as a circuit protective conductor or not, should be connected to the main earthing terminal. PVC and all flexible conduit should contain separate circuit protective conductors. A visual inspection can usually confirm this.

Trunking, both in steel and PVC, should be adequately supported throughout its length. Trunking is often fabricated on site and checks need to be made to see if cut edges have been sleeved where cables come into contact with them. As steel trunking has to be electrically continuous, bonding straps may have been required across each joint. These need to be checked to ensure that they are tight. As with conduit the trunking installation should be complete. End caps and lids should be in place and spare holes blanked off.

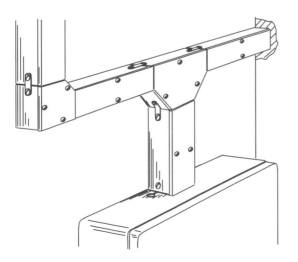

Figure 7.5   *All trunking lids should be in place and coupling links tight.*

Switchgear

Switchgear can be visually checked for damage, corrosion and to see if barriers are in place. However, before protective devices can be checked, the documentation stating the rating and type of device has to be consulted. It is important to know not only the rating of the device but also the type. For example when BS 88 fuses are used, the characteristics of fuses protecting circuits which include the need for the starting of motors will be different fuses of the same current rating protecting general service circuits.

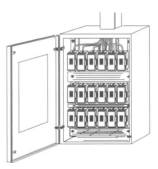

Figure 7.6   *Fuses may need to be checked for the correct type as well as rating.*

Electrical connections need to be checked for tightness and signs of overheating should always be investigated as this may indicate loose connections or overloaded cables.

Accessories

Accessories may contain loose connections as often a number of conductors have to be connected into one terminal. A check needs to be made for bad terminations and for signs of possible overheating. In addition to the current carrying conductors the cpc connections should also be checked to see if they are complete between mounting boxes and earth terminals. Damage due to mechanical pressure on the cable insulation should also be considered.

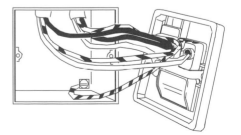

Figure 7.7   *Connections should be checked for electrical and mechanical soundness.*

Earthing and bonding

The earthing arrangements at the main intake should be checked to see if it is complete and correctly labelled and all connections and identification labels should be checked. Each circuit protective conductor should be correctly identified by colour coding, and terminated relative to the circuits with which they are associated. CPCs also need to be inspected to ensure bare conductors are not left.

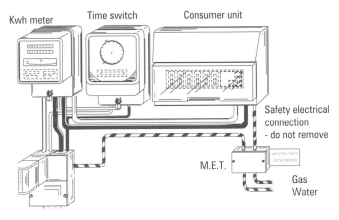

Figure 7.8   Domestic TN-C-S system - main earthing arrangement

In order to use earthed equipotential bonding and automatic disconnection of the supply as our means of protection against indirect contact we need to create an equipotential zone. This requires main equipotential bonding be carried out to all incoming internal services. A visual inspection may need to be made to examine connections to gas, water, structural steel, lightning protection systems and the like. This inspection should include checks on the use of the correct size, identification and labelling of conductors.

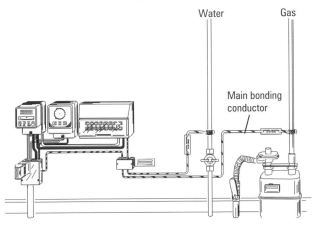

Figure 7.9   Visual inspection should be carried out to confirm connections are satisfactory.

Supplementary bonding is required in areas of increased shock risk, such as the domestic bathroom and checks need to be made for good connections and correct identification.

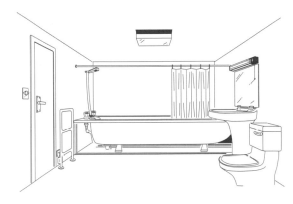

Figure 7.10   All exposed metalwork should be at the same potential.
Visual inspections should be made at all connections.

Residual current devices (RCDs) may be used to provide protection against direct or indirect contact where the conditions are particularly onerous or conventional methods cannot achieve the level of protection required. We shall consider the use of RCDs for protection against both forms of contact beginning with indirect contact. During the course of the inspection we should ensure that the appropriate devices have been installed and that they have the correct ratings and settings. It may be possible, at the inspection stage, to check the mechanical operation of the device by operating the test button. The performance and confirmation of protection by these devices will need to be confirmed as part of the testing process.

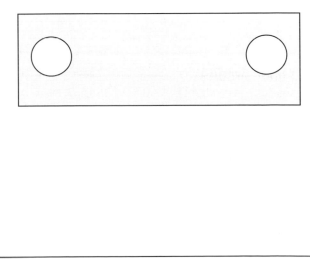

Type of circuit

Inspections also have to relate to the type of circuits used including:

* lighting circuits
* ring final circuits
* bathrooms
* alarm circuits

All circuits require a check to be made to ensure that the correct type and rating of protective device has been installed in accordance with the design. We also need to check that the cable installed is of the correct c.s.a. and current rating for the circuit, bearing in mind the installation method and other relevant factors, such as ambient temperature, grouping, thermal insulation and type of protective device used. However certain types of circuits or particular locations have additional requirements which should be considered and we shall consider some of these here.

Lighting circuits

Luminaires should be checked to ensure that they are suitable for the environment in which they are installed, are correctly fitted and, where appropriate, incorporate a suitable size and type of flexible cord. Enclosed fittings, or those where high temperatures are likely to occur due to their operation, should be either supplied in heat resisting cable or have a heat resisting sleeving fitted to the conductors at the terminations.

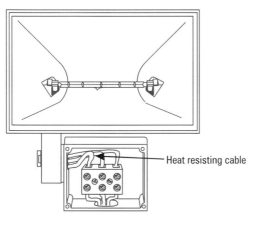

Figure 7.11   Where high temperatures are experienced, heat resisting cable must be used.

Ring final circuits

Ring final circuits with socket outlets to BS 1363, are different to other circuits insomuch as they are wired using a conductor whose current carrying capacity is less than the rating of the protective device. This is only possible because the circuit is wired as a ring and effectively has two cables to each point on the circuit, the actual distribution of current around the ring circuit is quite complex and will not be covered in this work book. Suffice it to say that BS 7671 does acknowledge the installation of this type of circuit arrangement as being acceptable providing it is a true ring circuit.

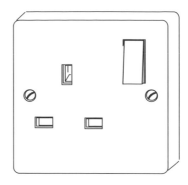

Figure 7.12   13 A socket outlet to BS 1363

If the ring is incomplete, has "multiple loops", that is a ring within a ring, or has excessive socket outlets supplied via spurs, then currents can flow which exceed the rating of the cables but not that of the protective device. This could result in overheating and damage to the cables and may result in a fire.

During the inspection process we may be able to confirm, visually, that the circuit is a ring where the circuit cables are visible throughout. Where this is not possible, we will need to undertake tests to confirm that the circuit is a true ring. The tests will be considered in the testing section of this workbook.

Bathrooms

Bathrooms are included in the "Special Locations" in Part 6 of BS 7671 and as virtually every domestic installation has a room containing a fixed bath or shower, are probably the most common locations of increased risk encountered in electrical installation work. There are special considerations that have to be made for the type and siting of equipment and accessories within bathrooms and these should be checked during the inspection process.

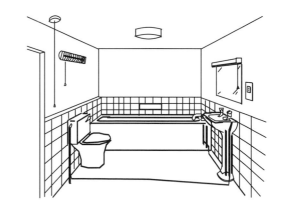

Figure 7.13   Equipotential bonding leads and connections must be checked.

In addition there are particular requirements for supplementary bonding between exposed and extraneous conductive parts within a room containing a fixed bath or shower. The terminations for the supplementary equipotential bonding conductors should be accessible for inspection and testing, so their presence and location should be confirmed during the inspection process.

Try this

Draw up a list of items that need to be visually inspected for each of the following situations.

1. A toilet area in a sports centre.

2. A reception area for a dental surgery.

3. A boiler room.

4. A large open plan office area.

In order to produce a quality installation we must first establish what standards are applicable to the type of installation being undertaken and ensure that we are familiar with the particular requirements.

Having done so, a continual monitoring process needs to be implemented consisting of regular inspection and testing with instruments. When tests are carried out, the results should be recorded and compared with the minimum or maximum acceptable values. Should any discrepancies occur further investigation will be necessary to establish the cause and take suitable remedial action in order not to compromise the overall installation. Without these regular checks the installation may be completed before any problems are identified and remedial work may be costly and time consuming.

The final inspection and testing undertaken before the installation is put into service should be confirmation that the installation meets all the required standards. The details of the installation and test results recorded demonstrate the condition of the installation prior to energising and immediately after energising and before the installation is put into service.

Inspection is an ongoing activity which begins when the installation work commences and continues through completion and then at regular intervals throughout the working life of the completed installation. Many faults can be prevented from developing into dangers if they are identified at an early stage. The inspection process should include all aspects of the installation and include the cables, conductors, accessories and equipment as well as any enclosures and control and protection devices. This applies to all the circuits which go to make the installation as a whole and to any altered or additional circuits added at a later date.

Self-assessment short answer question

1. Explain why old installations that have to be inspected and tested have to be approached differently to new installations.

Use the drawing for the factory in the Appendix.

1. Draw up a list of the items that need to be visually inspected for each of the following situations.

   a. Toilet area – ground floor

   b. Boiler room

   c. Reception area

   d. First floor office area

2. On the form provided in the Appendix "Schedule of Electrical Circuits" make a list of all circuits and possible protection devices for the factory in the Appendix.

8

Testing

Check that you can remember the following facts from the previous chapter.

Give an example of how each of the following senses may be used in an inspection of an installation:

smell –

touch –

hearing –

sight –

An inspection verifies that the _____ and _____ installed is to the specified standard.

On completion of this chapter you should be able to:

- ◆ describe how to carry out a continuity of protective conductor test
- ◆ describe at least one method of carrying out a verification of ring final circuit continuity test
- ◆ select the correct specification for an insulation test instrument
- ◆ describe how to carry out an insulation resistance test
- ◆ recognise the need for safety when carrying out an insulation resistance test
- ◆ identify equipment designed to give protection by electrical separation
- ◆ describe the use of barriers and enclosures to IP2X and IP4X
- ◆ describe how to carry out a polarity test
- ◆ recognise the need to carry out earth fault loop impedance tests
- ◆ describe the earth fault loop path
- ◆ describe how to carry out an earth electrode resistance test
- ◆ recognise the need to test residual current devices
- ◆ identify fixed equipment which may require separate insulation resistance tests
- ◆ recognise the need for testing portable electrical equipment
- ◆ describe how to carry out tests on portable equipment
- ◆ describe how tests can be carried out on certain items of electronic equipment

Part 1

Throughout this section reference is made to the recording the results of the tests we undertake on the installation. Whilst BS 7671 makes reference to the recording of the results of tests undertaken there is no document to indicate how this should be achieved. In Part 5 of Guidance Note 3, Inspection and Testing, published by the IEE, there are examples of forms of certification for use with electrical installations. Amongst these are schedules of test results and items inspected. Similar forms are available from trades associations, instrument manufacturers and independent publication companies which may be used for the certifying and reporting on electrical installations. Some contractors produce their own forms for certifying and all these are acceptable providing they contain at least all the information required by BS 7671.

Tip

When carrying out the inspection and testing on site, it is a good idea to have a copy of the schedules on which to record the results which can then be used to transfer the information to the final form for issue to the client.

There are two principal parts to the testing of an electrical installation or circuit;
- tests which are carried out to ensure the installation is safe to be energised, the "dead tests", and
- the tests carried out once the installation is energised to make sure it is safe to put into service, the "live tests".

When we are considering a brand-new installation the sequence of testing is started during the construction and followed through to completion of the installation. Once we are satisfied that the "dead tests" indicate the installation is safe to energise we may connect the supply and carry out the "live tests", checking the results to ensure the installation is safe to put into service.

When we are carrying out an inspection and test of an installation which has previously been energised and in use we need to ensure that the supply is switched off and the installation or circuit is suitably isolated before we carry out the "dead testing".

When carrying out testing on an electrical installation the tests should be undertaken in a particular sequence, which ensures that should the installation fail any one test it does not affect the tests already carried out. The IEE has included in their Guidance Note 3, Inspection and Testing, details of the sequence of testing. There are two sequences provided, one for the testing of new installations, the other for periodic testing of installations which have already been in service. The sequences are different to make allowance for the conditions which exist and in each case the sequence should be followed. Simplified versions of the two lists are produced below for ease of reference, but it would be advisable to consult Guidance Note 3 to consider all the information provided within that document.

Sequence of tests for new installations

The following tests should be carried out before the supply is connected (or disconnected as appropriate):

- continuity of protective conductors, main and supplementary bonding
- continuity of ring final circuit conductors
- insulation resistance
- site applied insulation
- protection by separation of circuits
- protection by barriers or enclosures provided during erection
- insulation of non-conducting floors and walls
- polarity
- earth electrode resistance

Once the electrical supply has been connected, recheck the polarity before conducting further tests

- prospective fault current
- earth fault loop impedance
- residual current operated devices

Periodic testing – sequence of tests

- continuity of all protective conductors
- polarity
- earth fault loop impedance
- insulation resistance
- operation of devices for isolation and switching
- operation of residual current devices
- prospective fault currents

Other tests may also be undertaken where appropriate.

For the purpose of this study book we shall consider the tests required following the sequence required for a new installation.

Continuity tests

The first tests in the sequence are concerned with confirming the continuity of the protective conductors and ring final circuits. It is important that the continuity of these conductors is confirmed before any further tests are undertaken and we shall consider the tests for each of the protective conductor types and the ring circuit continuity under this general heading.

Continuity test instruments

To carry out the tests for continuity we require an Ohmmeter which has a low ohms range or, as most electricians use, an insulation and continuity test instrument set on the continuity range. Remember we are testing the continuity of conductors intended to carry current and therefore we are likely to obtain fairly low readings, often less than 1 Ω. There are recommendations made in European standards that instruments used for continuity testing should meet the following criteria,

1) an open circuit output voltage of 4 to 24 V
2) a minimum short circuit current of 200 mA (either a.c. or d.c.)

and it would be advisable to ensure that any instrument purchased to carry out continuity tests is manufactured to meet these requirements.

As we are testing quite low values of resistance, it is important that we do not include the resistance of the test leads in our results. We can do this in two ways, the first being to measure the resistance of any test leads before we carry out any tests and then subtract the lead resistance from the result of each test. Alternatively some instruments have the facility to zero the instrument reading to compensate for the resistance of the test leads. The test leads are shorted together and the instrument is set to read zero. Once this has been done, the instrument will automatically read the true resistance of the conductor being measured providing the same leads are used and the instrument is not switched off. The process will need to be repeated each time the instrument is switched on or different leads are used.

Continuity of protective conductors

In order to test the continuity of a protective conductor we must ensure that there are no parallel paths connected which would give a false indication of continuity. For this reason each protective conductor should be tested before it is finally connected to the rest of the installation. On a new installation this would be before the final connection to the distribution board or main earthing terminal. The purpose of the test is to ensure that the conductors are continuous throughout their length and that there are no high resistance joints or connections.

Continuity of earthing and main equipotential bonding conductors

To test the continuity of earthing and bonding conductors we need to employ the use of a long wander lead to test between the main earthing terminal position and the termination at the point where the services enter the building, as these are not usually adjacent to each another. Once the termination has been made to the incoming service we can run out our long lead and carry out the continuity test before we make our final connection to the main earthing terminal.

Remember
Always deduct the resistance of any test leads from the overall result to get a true continuity test value.

Example

Example
Test of main equipotential bonding conductor between the main earthing terminal and the incoming water service.

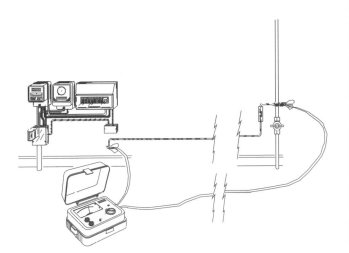

Figure 8.1

Main earthing terminal termination to incoming water service test value = 0.5 Ω

Resistance of the test leads = 0.45 Ω

Resistance of the main bonding conductor
= 0.5 − 0.45 Ω = 0.05 Ω

We can verify this value by using the conductor resistance values given in IEE Guidance Note 1, and this gives a resistance of 1.83 mΩ per metre for a 10 mm² copper conductor. The length of the main bonding conductor should therefore be in the region of 0.05 ÷ 0.00183 = 27.3 m and we can verify this by a quick on-site measure.

The requirement to confirm the continuity of protective conductors means we also need to test any supplementary bonding conductors to ensure their continuity. These conductors need to be connected at the point they are installed as they are not connected directly to the main earthing terminal.

Remember
Where supplementary bonding conductors are installed to provide protection against indirect contact the continuity of the supplementary bonds and their connections must be carried out.

Continuity of circuit protective conductors

We can carry out the test for continuity of circuit protective conductors in the same way as we did the main bonding conductors, testing between the cpc at the distribution board and the earthing terminal(s) on the final circuits. However the use of long test leads and running them around the building is both time consuming and a potential tripping hazard for people in the building, and can be particularly awkward in large buildings. The alternative method for this test is to use the associated phase conductor in place of the "wander lead". This method of testing the continuity of the cpc is generally known as the "$R_1 + R_2$ method" as it involves the phase conductor resistance, R_1, and the cpc resistance R_2. We can then record the values for $R_1 + R_2$ onto the schedule of test results. If this method is adopted there is no need to record the value of R_2 separately on the schedule.

CPC continuity using the "$R_1 + R_2$ method"

Whilst this is not essential to establish the continuity of the protective conductor, it is a convenient method and fulfills three of our test requirements in a single operation.

- Confirms continuity of the protective conductor
- Confirms the correct polarity of the circuit under test
- Provides the combined resistance of the circuit earth fault path, this will be required if we are unable to carry out a full earth fault loop impedance test, for example where an RCD is fitted to protect the entire installation.

The method of carrying out this test involves connecting the phase and cpc together, either at the distribution board or at the furthest point of the circuit from the distribution board. The instrument is then connected across the phase and cpc at the other end of the circuit under test.

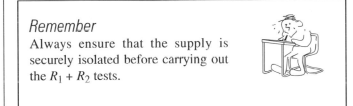

Remember
Always ensure that the supply is securely isolated before carrying out the $R_1 + R_2$ tests.

For the purpose of this explanation we shall consider the use of a shorting link connected at the consumer unit and the test being carried out at the furthest point of the circuit under test. The link used for the connection between phase and cpc should be as short as possible in order to minimise the effect on the readings, where possible the instrument should be "zeroed out" to include the link.

Ensure the supply is securely isolated and insert the shorting link between the phase and the cpc for the circuit to be tested, as shown in Figure 8.2.

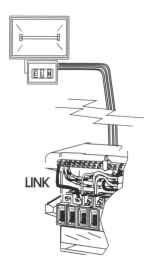

Figure 8.2   *The link is in place in order that the resistance of the phase and circuit protective conductors can be measured.*

The instrument can now be connected between the phase and cpc at any point on the circuit and a reading obtained. However, we are not only concerned with the not only continuity of the cpc but also the "worst case" ie, where the value of $R_1 + R_2$ is at a maximum and this is generally at the end of the circuit furthest from the distribution board. The instrument is connected using the crocodile style clips or the probes, whichever is the most convenient, as shown in Figure 8.2.

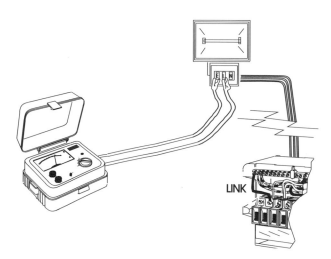

Figure 8.3   *With the link in place the resistance of the cpc can be measured at any point on the circuit.*

Once we are sure that any functional switches on the fixed wiring are closed the test is carried out and reading obtained is recorded and entered on the schedule of test results.

Where a circuit supplies a number of points we can ensure that we have continuity of cpc and correct polarity at each point by carrying out a similar test at each point. The value recorded on the schedule of test results is the highest value obtained, which should be that at the end of the circuit furthest from the distribution board.

Once the continuity testing of the circuit is complete the shorting link is removed and the circuit reinstated.

Remember it is important to test the circuits individually as we must make sure that there are no parallel return paths whilst the test is carried out as the aim is to confirm the continuity of the protective conductor for the circuit and not any fortuitous connections which may not be reliable under fault conditions.

Testing of steel cpcs

Where the cpc for a circuit is provided by a steel enclosure, such as a metal conduit or trunking, the continuity test should be carried out initially by the use of the wander lead or $R_1 + R_2$ method. Should the results or the physical condition of the enclosure give cause for concern over the suitability of the cpc, then a further High Current Test should be carried out. This would need to done using a high current low impedance ohmmeter. This will require a supply as the current used for this test is recommended as 1.5 times the design current of the circuit up to a maximum of 25 A, at 50 V. This test is not usually required on a new installation.

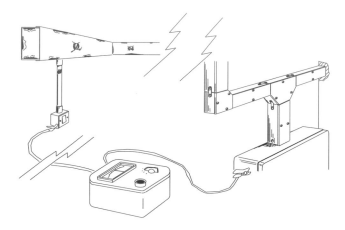

Figure 8.4   *A high current test may be required where steel is used as the circuit protective conductor.*

Part 2

Continuity of ring final circuit conductors

The ring final circuit has been used extensively for the provision of general purpose power sockets complying with BS 1363, better known as the 13 A socket. However, ring circuits may be used to supply other equipment to overcome particular problems with an installation, location or safety consideration. The method of testing for continuity is applicable to any ring circuit but for ease of explanation we shall be considering a typical BS 1363 ring circuit.

As we discovered earlier in this workbook, ring final circuits are different to other circuits insomuch as they are wired using a conductor whose current carrying capacity is less than the rating of the protective device. If the ring is incomplete, has "multiple loops", that is a ring within a ring, or has excessive socket outlets supplied via spurs, then currents can flow which exceed the rating of the cables but not that of the protective device. This could result in overheating and damage to the cables and may result in a fire.

Unfortunately it is relatively easy to make a mistake when installing ring circuits, particularly when using single core cables in conduit or trunking. In order to confirm that the circuit is a true ring circuit we need to carry out a sequence of tests on each ring circuit.

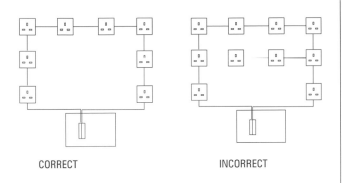

CORRECT   INCORRECT

Figure 8.5   *Line diagrams showing correct and incorrect connections of a ring circuit.*

First we need to ensure that the circuit is securely isolated, all the socket outlets have been installed, no equipment or appliances are connected and disconnect the cables to the ring circuit at the distribution board. This can be done at the origin of the ring in the distribution board, or at a socket outlet close to the distribution board, whichever is the most convenient. As we are considering the initial verification of ring circuit continuity we shall carry out the testing from the distribution board.

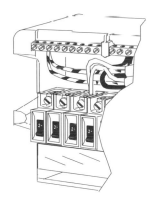

Figure 8.6   *Ring final circuit connected to a consumer unit before tests commence.*
(For clarity the cables to other circuits have not been shown.)

Tip
When carrying out a new installation it is often helpful to complete the installation of the socket outlets to the ring circuit and then carry out the ring continuity tests as the circuit is terminated at the distribution board.

The instrument we use must be a low impedance ohmmeter or continuity tester, which is capable of measuring small variations in resistance. As we are measuring conductors the values will be quite low and the variations may be small and so the instrument should be able to register values in the order of $0.05\ \Omega$.

The first stage of the test is to confirm the conductors are connected in a ring and to establish the resistance of each conductor from end to end. As the ring circuit begins and ends at the same protective device, we simply connect the instrument between the two phase conductors and test to obtain a reading. Remember the resistance of the leads will need to be either zeroed out or deducted from each reading obtained. We shall assume the instrument has been zeroed to take account of the leads. As we are considering the standard ring circuit, the phase conductor will be a 2.5 mm^2 copper conductor which has a resistance in the order of $0.007\ \Omega$ per metre. When we test the phase to phase reading the result, divided by 0.007, will give an approximate total cable length for the ring.

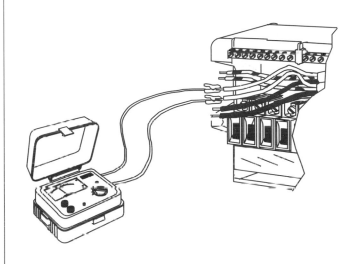

Figure 8.7   *Conductors disconnected from the consumer unit and an ohmmeter connected across the two phase conductors of the ring circuit.*

So if our test result phase to phase on the ring circuit is say, $0.53\ \Omega$ then we can estimate the total length of the conductor by using the formula length = total resistance ÷ resistance per metre, in this case $0.53 \div 0.007 = 75.7$ metres. This length is an approximate value which gives us some idea of the total length of conductor involved.

We now repeat the test with the two neutral conductors and the results should be substantially the same, as the cables should follow the same routes, say within $0.05\ \Omega$.

The final part of the first stage is to repeat the test with the cpc and where the installation is carried out using conventional twin and cpc cables the cpc is generally one size smaller than the live conductors. In this case we shall assume this to be the case and therefore the total resistance will be proportionally higher and as a rough guide it would be somewhere in the order of 1.667 times higher. This value is arrived at by using the proportion of the c.s.a. and as resistance is directly proportional to c.s.a. we can establish the ratio by 2.5 (mm^2) ÷ 1.5 (mm^2) = 1.67. In which case the resistance end to end of the cpc in our example would be in the order of $0.53 \times 1.67 = 0.885 \, \Omega$.

If the values are not within an acceptable tolerance then the circuit requires some additional investigation to establish the reason, remember that the conductors should follow the same routes and be of approximately equal lengths. If a high resistance is found or an open circuit is established on any of the conductors then, again, further investigation is needed.

Tip
Where more than one ring circuit is installed, the first check in the event of an open circuit is to ensure that the two conductors are for the same circuit and a test of continuity to the other ring circuit conductors will generally confirm whether this is indeed the case.

Remember
Where the cpc comprises a steel enclosure, we are unable to carry out the ring circuit test on the cpc but remember that we should have already checked for cpc continuity. The confirmation of ring circuit continuity for the live conductors must still be carried out and this is done in the same way irrespective of the nature of the cpc.

We need to record the values we obtain when carrying out this initial test as we shall need them later when confirming the ring circuit continuity. The values are generally recorded as

r_1 = phase to phase resistance
r_2 = cpc to cpc resistance
r_n = neutral to neutral resistance

Providing the tests have confirmed that the conductors have been correctly identified we can proceed with the continuity testing.

The next stage is to "cross connect" the phase and neutral conductors, that is, connect the phase of one end of the ring circuit to the neutral of the other end and vice versa as shown in Figure 8.8. This is relatively easy when the installation is in sheathed cables but may be a little more difficult when using singles in trunking and conduit.

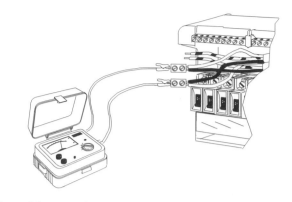

Figure 8.8 *The phase of one cable is connected to the neutral of the other, and this is repeated for the other phase and neutral.*

The resistance of the conductors is then measured across the connected pairs as shown in Figure 8.8 and the values obtained are recorded. These values should be approximately half the value obtained for r_1 or when we tested phase and neutral conductors end to end. Now, using a plug top, we can test at each socket outlet between the phase and neutral conductors, with the cross links in place, as shown in Figure 8.9.

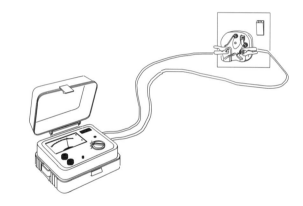

Figure 8.9

Remember
If this test is carried out from a socket outlet, we must also check the reading at the distribution board.

The test readings obtained at each socket and the distribution board should be substantially the same as the reading taken from the test in Figure 8.8. Socket outlets connected via a spur from the ring circuit will give slightly higher values of resistance, and the precise increase will be proportional to the length of the branch cable. If we find that the resistance values increase as we move further away from the distribution board this indicates that we have not cross connected the ring circuit but have connected the phase and neutral conductors of the same ends together. In such cases the ring circuit has to be cross connected correctly and the test carried out to confirm ring circuit continuity and the readings should now be substantially the same at each point.

We must now repeat the process only this time we cross connect the phase and for the ring circuit. The value of resistance will be higher than in the previous test if the cpc is of a smaller c.s.a. than the live conductors. Again we carry out the test at each outlet, between phase and cpc this time, and we should expect some variance in the values. As before the values obtained at sockets connected via a spur will produce higher values, proportional to the length of the branch cable.

The highest value obtained during this test process represents the $R_1 + R_2$ value for the ring circuit and should be recorded on the schedule of test results in the $R_1 + R_2$ column. The value obtained should be in the order of $(r_1 + r_2) \div 4$, using the $r_1 + r_2$ values obtained in the first test. This is due to the circuit now comprising of the two conductors connected in parallel and, because they are connected as a ring circuit, the effect length of the conductor is also halved to any point on the circuit, compared to the end to end length. The total resistance value is therefore halved as the effective c.s.a. is doubled and halved again as the effective length is halved, hence a quarter of the original $r_1 + r_2$ value.

Example 1

At the distribution board as in Figure 8.8

Phase to phase	= 0.8 Ω
Neutral to neutral	= 0.8 Ω

At the distribution board as in Figure 8.8

$$= 0.4 \ \Omega$$

At sockets on the ring when connected as in Figure 8.8

Phase to neutral	= 0.4 Ω

If all of the sockets are part of a single ring then the reading at each outlet should be substantially the same.

Example 2

For this example we will consider a 25 m ring final circuit wired in 2.5 mm^2 with a cpc of 1.5 mm^2.

Phase to phase	= 0.4 Ω
cpc to cpc	= 0.6 Ω

At the distribution board, or sockets on the ring, when connected as in Figure 8.9 the resistance should be close to

$$= \frac{0.4 \times 0.6}{0.4 + 0.6} \ \Omega$$

(product over sum)

$$= 0.24 \ \Omega$$

Note:

Take care if single-core cables are used that OPPOSITE ends of the phase and neutral conductors are bridged together.

Once we have completed the test, we can remove the cross connections and terminate the ring circuit to the distribution board. We have, by this test process, confirmed the circuit to be a ring, hence the cables will not be subject to overload, confirmed earth continuity of the cpc to each point on the circuit and tested the correct polarity of each socket, three tests in the one process.

Points to remember

The first tests undertaken are to confirm the _____ of conductors, including protective conductors and ring circuit conductors.

It is essential that all extraneous conducting parts of main incoming services, such as _____, _____ and oil, are connected to the main earthing terminal. Tests must be undertaken to confirm the continuity of these main equipotential bonding conductors.

Tests must be undertaken to confirm the continuity of supplementary bonding conductors in areas of increased shock risk.

Tests must be undertaken to confirm the continuity of protective conductors.

Ring circuits which are not correctly connected can become a _____ hazard.

All continuity tests should be undertaken using a _____ _____ ohmmeter or continuity tester.

It may be necessary to carry out a high current continuity test on _____ enclosures used as protective conductors if the continuity established by conventional continuity testing appears to be suspect.

Short answer questions

All questions in this section refer to the specification and plans of the small factory shown in the Appendix.

1. The ring final circuit in the first floor office area is supplied from the consumer unit on the first floor landing. Explain how a continuity test for this ring final circuit could be carried out.

2. The steel conduit through to the light fitting(s) in the works office has to be tested to check it is suitable as a circuit protective conductor. List the test equipment suitable for the test and any special considerations which need to be made.

Part 3

Insulation resistance testing

In the previous tests we were concerned with establishing that a good electrical connection was made by recognised conductors and that the resistance of these conductors was sufficiently low for high current to flow in the event of a fault occurring. When we carry out insulation resistance testing, we are testing the resistance of the insulation separating live parts, conductors in particular, from each other and from earth. So this time we shall be looking for high values of resistance and the instrument used must be capable of measuring high values of resistance, in the range of $M\Omega$.

We are also testing to establish that the insulation is going to be able to withstand the rigours of everyday operation and the voltages likely to be encountered. To test this we use a voltage higher than that to which the insulation will normally be subjected and the test voltage required will be dependent upon the normal operating voltage of the circuit to be tested. Advice on the applied voltage for this test is given in Table 71A of BS 7671: 1992 and by reference to this we can see that for circuits other than SELV and PELV and those operating above 500 V the test voltage is 500 V.

> ### Remember
> The resistance of a cable conductor is inversely proportional to its c.s.a. and proportional to its length (it gets smaller as the c.s.a. increases and higher as the length increases).
>
> The resistance of a cable insulator is directly proportional to its c.s.a. and inversely proportional to its length (it gets higher as the thickness increases and lower as its length increases)
>
> When carrying out the insulation resistance tests the c.s.a. of conductors and that of their insulation is generally fixed and the only variable is the length.
>
> When several cables are connected in parallel, such as at a distribution board, then the overall effect is to connect a number of resistances in parallel, so the overall effect is for a total resistance less than that of any individual resistance.

Insulation resistance test instruments

The instrument used to carry out the insulation resistance test must be able to meet the requirements of BS 7671 for the insulation being tested. In this case we are considering installations operating at 230/400 V, so our test instrument must be capable of producing a test voltage of 500 V. However, as we can see from Table 71A in BS 7671, there are separate criteria for circuits operating at other voltages and instruments used for testing these circuits must meet the appropriate requirements.

The reason for using a higher voltage level is because they are sufficiently high to reveal any breakdown or weakness in the insulation, which is "voltage sensitive". In addition the instruments must be able to supply the test voltage when loaded to 1 mA. This is generally achieved by the use of an instrument with a battery power source and electronic circuitry to produce the required output and may have either an analogue or digital display.

Figure 8.10   Digital insulation resistance tester

Figure 8.11   Analogue insulation resistance tester

Safety whilst carrying out the insulation resistance tests

There are two main aspects to the safe implementation of the insulation resistance test.

The **FIRST** is that the test should be carried out on circuits disconnected from the supply. To assist the test engineer some instrument manufacturers have built voltage indicators into the equipment. If this indicator is activated then no tests should be carried out until the supply has been isolated. Where test equipment does not have this facility, the circuits being tested should be checked to confirm there are no supplies before the insulation resistance tests are carried out. Where circuits contain capacitors the test equipment may charge up the capacitor and care will need to be taken to avoid discharging the capacitor through your body.

The **SECOND** safety consideration that should always be given serious thought is that the test voltages being injected into the installation are high and could cause serious accidents. Precautions should be taken to ensure nobody can become part of the circuit when the test is being carried out.

Before we can carry out our insulation resistance tests there are a number of procedures that must be carried out.

These include
- ensuring the supply is securely isolated and that there is no supply to the circuits to be tested
- removing all lamps
- disconnecting all equipment that would normally be in use
- disconnecting and bypassing any electronic equipment that would be damaged by the high voltage test, (this may include lamp dimmer switches)
- putting all fuses in place
- putting all switches in the ON position (unless they protect equipment that cannot otherwise be disconnected)
- testing two way or two way and intermediate switched circuits with the switches in each direction unless they are bridged across during the testing

Once we have satisfied ourselves that the above conditions have been complied with, we can begin testing for insulation resistance.

Insulation resistance test on a new installation

We can now begin to undertake the insulation resistance testing on our installation. We shall first consider the process for testing a new domestic installation and then consider the implications of a larger installation, single circuit tests and testing on installations which have previously been energised and in use.

New domestic installation

When testing a new domestic installation this can be carried out from the tails to the consumer unit before they are connected to the supply. Figure 8.12

Figure 8.12   A domestic consumer unit ready for connection to the supply.

In this case the main switch or switches must be on, all fuses complete and in place, or all circuit breakers switched on.

The first test is between phase and neutral as shown in Figure 8.13.

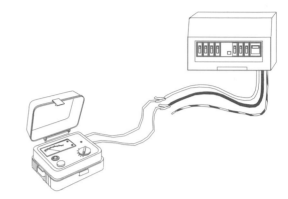

Figure 8.13   An insulation test instrument connected to the tails of a domestic consumer unit. All switches in the unit are in the ON position.

The resistance for this should be greater than $0.5 \, M\Omega$. If it is less than this further investigation must follow. This can be started by repeating the test with each circuit isolated in turn. When the circuit with the fault has been identified, a further breakdown of this circuit can be made. Often a lamp has been left in by mistake or the immersion heater is on. However until there is a satisfactory explanation for this result no further tests should be carried out.

When a satisfactory result has been obtained for the first test the phase and neutral can be connected together and the test instrument connected between these and the main circuit protective conductor. As before this resistance must be equal to or greater than $0.5 \, M\Omega$. If a meter with an analogue scale is being used, the needle may point to INF or ∞. This means that the reading is greater than the range of the instrument and is not something that can be recorded on a report. If the maximum reading on the meter scale is $100 \, M\Omega$ then an infinity reading should be recorded as greater than $100 \, M\Omega$ or $> 100 \, M\Omega$.

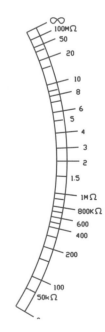

Figure 8.14   The scale of an analogue meter showing "0" at one end and "∞" at the other.

Insulation resistance testing on larger installations

In theory there is very little difference between carrying out an insulation resistance test on a small or large installation. In practice however, there are a number of extra considerations that must be given.

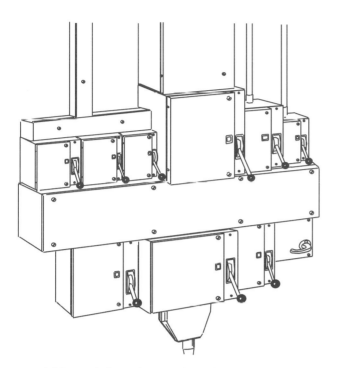

Figure 8.15   A three-phase intake with TP & N bus-bar unit and switch fuses.

If all of the preparation was carried out insomuch as the lamps were removed, switches were all on and so on, and a Megohmmeter was connected across the supply cables of a new large installation before the supply was connected, the chances are that the results would not be acceptable. This would usually be due to the enormous total length of cable used in the installation. As the meter is trying to measure insulation resistance and not conductor resistance, the longer the total length of insulation the lower the total insulation resistance would be.

Allowance for this is made in BS 7671 and we can break the installation down into the main switchboard and each distribution circuit tested separately with all its final circuits connected. This will allow us to undertake a test at each sub-distribution board with all the circuits connected and a test at the main distribution position with all the distribution circuits connected but the sub-distribution boards isolated. The principle of this breakdown is illustrated in Figure 8.16 and at each point the total value of insulation resistance must be no less than 0.5 MΩ.

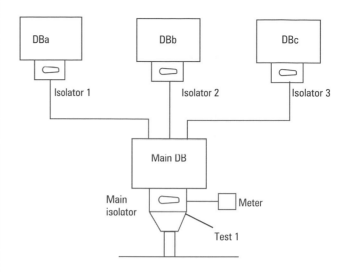

Figure 8.16   Test 1 at the main isolator/Distribution board with Isolators a, b, and c in the off position

Test 2 at Isolator 1 outgoing terminals with all DB(a) fuses in, lamps removed etc.

Test 3 at Isolator 2 outgoing terminals with all DB(b) fuses in, lamps removed etc.

Test 4 at Isolator 3 outgoing terminals with all DB(c) fuses in, lamps removed etc.

Industrial and commercial installations would usually be designed for use with a three-phase four wire supply. The first test in this case would be between phases. The minimum acceptable insulation resistance is the same as domestic installations, i.e. 0.5 MΩ. After the insulation between phases has been proved to be satisfactory, the three phases can be connected together and the insulation resistance measured between them and the neutral conductor. Once this has been proved to be acceptable, the neutral can be added to the phases and the resistance measured between them and the circuit protective conductors. This again must not be less than 0.5 MΩ. This series of tests may have to be repeated many times to complete the whole installation.

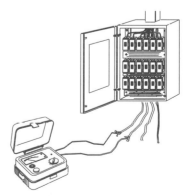

Figure 8.17   An insulation resistance test carried out on a three-phase distribution board.

Single circuit testing

When an alteration or addition is carried out to an existing installation, we may need to test that circuit alone in order to compile our certification. We need to make the same check for equipment and isolation as we do for the installation and carry out the same procedure for testing the circuit as we would for an installation. We do need to be careful particularly if we are carrying out the test in the vicinity of energised circuits and equipment.

Older and in service installations

On older installations where tests are carried out to verify that the insulation is still in an acceptable condition, it may not be possible to carry out tests on the whole installation at the same time. In these cases each circuit can be tested separately and the results recorded. As equipment is more likely to be in use in existing installations more care has to be taken to ensure it has all been disconnected before the tests are carried out. The acceptable resistances are the same on old installations as on new.

On existing installations sections may have to be isolated and tested to meet the requirements of those using the installation. The important requirements are that
- all sections of the installation are tested
- a record is made of all test results and what they apply to any results that are not up to specification should be reported in writing to somebody in authority.

Site applied insulation

The next test in our sequence is for the insulation of site applied insulation. This involves the testing of insulation which is applied on site and relied upon for protection against direct contact. As the majority of equipment used has the insulation for protection against direct contact provided by the manufacturer of the equipment this test is not normally required. Site applied insulation does not apply to the construction of, say, a control panel using proprietary relays and switches enclosed in pre-made cabinet using single insulated cables. All the component parts have the insulation applied by the manufacturer before they are supplied to the wholesaler or contractor.

Where the site applied insulation is relied upon for protection against indirect contact then the test should confirm that the insulating enclosure provides a degree of protection equivalent to a least IP2X or IPXXB. These degrees of protection will be considered further in the requirements for protection by barriers and enclosures provided during erection, later in this chapter.

The test involves, besides the insulation resistance test, the application of a high voltage test, where the equipment is subject to a flashover or breakdown test at an equivalent voltage to that for a manufacturer's type test for similar equipment. As this test requires specialist equipment and knowledge, we shall not be considering the requirements in this workbook. Should you be required to have equipment tested in this was it would be advisable to contact a specialist in this field of work.

Protection by separation of circuits

This test is required where the method of protection is by the use of SELV or PELV and electrical separation. For these systems to be employed and provide protection against electric shock the source of the supply must provide separation from live parts of other systems. A simple example of electrical separation is an electric shaver socket manufactured to incorporate a BS 3535 isolating transformer. This, by nature of its construction, provides an output of 240 V which is electrically isolated from the supply completely. This item of equipment provides protection for the single outlet incorporated within it and uses electrical separation to provide protection against electric shock.

The protection by electrical separation requirements for circuits extend beyond this to incorporate the source of the supply and any associated wiring. We shall consider the requirements for the supply and the circuit conductors beginning with the source of the supply.

Sources of supply

There are a number of sources of supply detailed in BS 7671 and the most common, for general installation work, are those derived through a BS 3535 isolating transformer. It is usual for such sources of supply to be type tested by the manufacturer and provided it can be shown that the source has been type tested and complies with the appropriate standard then no further testing of the source will be necessary. If this is not possible then it will be necessary to carry out tests to confirm that the source of supply does meet the standard. As these tests are highly specialised, we shall not be considering them here and should they become necessary it would be advisable to contact an expert in this field.

Protection by electrical separation

On occasions it is required to isolate the supply electrically. An example of this is the use of BS3535 shaver sockets.

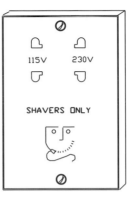

Figure 8.18   *A shaver socket to BS 3535 incorporating a double wound transformer with no connection to earth on the secondary winding.*

The electrical supply to these has no electrical connection to the output. The transformer used is an isolating type where there is no electrical connection to earth on the secondary side. Safety extra low voltage systems work on similar principles. Here a supply is provided that has no electrical connection on the secondary output that is referenced to any connection to earth.

To verify that the output is completely separate from earth an insulation test must be carried out. This test is carried out at 500 V d.c. for one minute and at the end of that period the resistance should not be less than 5 MΩ.

Protection by barriers or enclosures

So far all of the tests have involved the use of instruments to electrically test the soundness of the installation. It is however possible to have the situation where equipment can pass the electrical tests but still leave live parts exposed to touch.

In order to ensure this is not the case we need to carry out tests to ensure that there is no possibility of contact with live parts. Such risk may arise as a result from the construction of an item of equipment or enclosure on site. More often it is the result of modifications made to items of equipment, such as making entry holes into consumer units and accessories, in order to take our cables into them. There are two principal tests which need to be undertaken and both refer to the requirements identified within the IP code.

The first requirement is that all barriers and enclosures must meet the requirements of IP 2X or IP XXB, and this involves the use of the "standard finger" test probe, as shown in Figure 8.19.

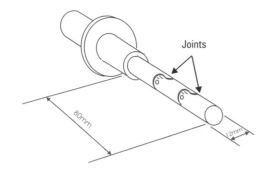

Figure 8.19   *A test probe to IP2.*
                  Joints may be used to simulate a finger.
                  For full details and dimension refer to BS 3042.

The "standard test finger" is capable of bending through 90°
twice, as a normal finger does, and is intended to establish
whether there is any possibility of contact with live parts
through the insertion of a finger, without the risk of electric
shock to the person carrying out the test. In terms of contact the
requirement of IP XXB is the same as IP 2X.

The second test is applied to the top surfaces of enclosures and
is more onerous than that required for IP 2X. For this test any
opening on the top surface of the enclosure must offer
protection against the entry of a wire or solid object larger than
1mm in diameter.

Should any of the openings fail to meet the requirements then
the opening must be reduced to meet the requirements.
Providing the openings are suitably sized, and any unused
openings are closed, a visual inspection will generally be
sufficient to ensure the installation meets the requirements.

Insulation resistance tests are carried out to verify that there is
no breakdown between phases, phases and neutral, and phases
and earth.

Points to remember ◀ – – – – – – – – – – – –

The tests must be carried out at twice the working voltage up to
500 V d.c.

Where electrical equipment is not connected when the circuit
tests are carried out it must be tested separately.

On occasions when equipment is built up on site, tests must be
carried out to ensure it conforms to the appropriate British
Standard.

It is possible to isolate an electrical supply using double wound
transformers. In these cases tests must be carried out to verify
there is no electrical connection between the primary and
secondary windings.

It is important to ensure that no live conductors of terminals are
left exposed to touch. Where barriers or enclosures have been
provided, other than with British Standards approved
equipment, tests must be carried out to confirm that probes to
IP2X and IP4X cannot penetrate through to live parts.

Self-assessment short answer questions

All questions in this section refer to the specification and plans
of the small factory shown in the Appendix.

1. An insulation resistance test is to be carried out on the
    main factory lighting circuits. Explain what preparations
    must take place before the tests can be carried out.
    Remember that the installation has been energised and is
    in use.

2. The lighting in the reception area has just been rewired.
    Describe, with the aid of a sketch, how insulation
    resistance tests should be carried out before the supply is
    reconnected.

3. A bench in the factory has electronic test equipment
    permanently wired into switched fused spur plates.
    Explain how an insulation test can be carried out on these
    circuits without damaging the electronic equipment.

4. A visual inspection reveals that an MICC cable in the
    boiler house has been dented. Explain how you would test
    this cable to ensure it is safe to continue using.

Part 4

Insulation of non-conducting walls and floors

The use of this method of protection is restricted to certain special locations, and the circumstances are such that the installation needs to be under effective supervision. As this type of installation is carried out by specialist contractors and requires specialist testing procedures we shall not consider them is this workbook. Further information and guidance on this subject is contained in IEE Guidance Note 3, Inspection and Testing.

Polarity testing

We have considered the testing of polarity when we were carrying out the $R_1 + R_2$ earlier in this workbook and it was stated that we could use this to confirm the polarity of circuits being tested. We do have to ensure that all the circuits are correctly connected and that all protective and control devices are connected in the phase conductors. In order to do so we shall consider the process necessary to confirm this which must be carried out **before** the installation is connected to the supply.

For safety reasons all switches, overcurrent protection devices and control contacts, must be in a non-earthed conductor, Figure 8.20.

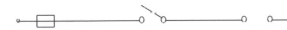

Figure 8.20   *All devices capable of breaking the circuit should be placed in the non-earthed conductor, i.e. the phase*

The only exceptions to this are where phase and neutral conductors are operated simultaneously with mechanical links coupling the operation.

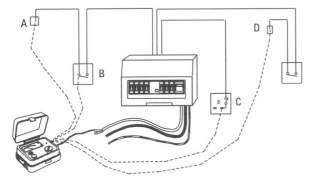

Figure 8.21   *Polarity tests are carried out to verify that*
A – the phase at the lamp holder is switched
B – the switch is in the phase conductor
C – the switch to the outlet is in the phase conductor
D – the centre pin of an ES lampholder is connected to the phase conductor

Polarity tests can be carried out using continuity test instruments. There is no requirement to record readings as this is a check to ensure that the installation connections have been made in the correct conductors.

On domestic installations the tests should verify that:
- all lighting switches are connected in the phase conductor
- the centre pin of ES is connected to the phase conductor
- the switches on switched socket outlets are connected in the phase conductor
- the correct pin of socket outlets is connected to the phase conductor
- where double pole switches are used, such as on immersion heaters, the phase and neutrals have not been swapped over

Industrial installations will include all those listed for domestic installations but will also include others due to the complexity of the installations. Most commercial and industrial installations are connected to a three-phase supply and include some three-phase loads. Control devices and isolators used in these circuits must switch the three phases simultaneously so that a single phase cannot be left connected.

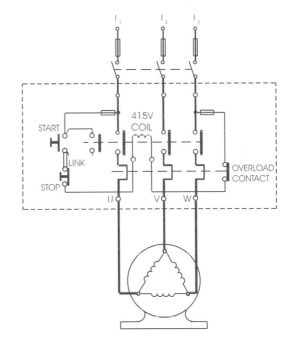

Figure 8.22   *Three-phase direct-on-line starter with a triple pole switch fuse*

Remember
Where it is impractical to use long leads going to each part of the building, the cpc of the circuit can be used as the return conductor. At the distribution board the cpc is connected or linked to the phase conductor. The continuity tester can now be taken to each outlet in turn and connected between the phase and cpc. A reading will indicate the correct polarity.

So far we have considered tests that need to be carried out on the installation to ensure it is in good condition and that it is safe for the supply to be connected. Once we are satisfied that the installation is safe to energise we can connect the supply but before the installation can be put into service we must carry out some further tests and checks. The following tests are carried out once the supply is available and the installation has been confirmed as safe to energise.

Earth fault loop impedance test

A designer, when calculating out an installation, should confirm at each stage that should a phase to earth fault develop the protection device will operate safely and before any permanent damage can occur in the installation. The time protection devices take to operate is directly related to the impedance of the earth fault path. Although the designer can calculate this in theory it is not until the installation is complete that the calculations can be checked.

Remember this test, unlike the others we have looked at so far, must be carried out when the installation is connected to the Supply Company's cables. The instrument used for this test is an Earth Fault Loop Impedance Test instrument, sometimes referred to as a "Loop Tester". Before starting to carry out these tests it is important to have documentation giving the maximum values for each circuit.

Try this
Using a copy of BS 7671:1992 complete the following tables:

Table 8.1 (from Tables 41B1 and 41B2) for 0.4 s disconnection with U_o = 230 V

Maximum Earth Fault Loop Impedance (Z_s)					
Fuses to BS88 Part 2 and Part 6					
Rating (amperes) 6	10	16	20	25	32
Z_s (ohms)					
Type 2 miniature circuit breakers to BS 3871					
Rating (amperes) 5	10	15	20	30	
Z_s (ohms)					

Table 8.2 (from Table 41D)

Maximum Earth Fault Loop Impedance (Z_s) for 5 sec disconnection time with U_o= 230 V					
Fuses to BS88 Part 2 and Part 6					
Rating (amperes) 6	10	16	20	25	32
Z_s (ohms)					
Fuses to BS 1361					
Rating (amperes) 5	15	20	30		
Z_s (ohms)					

Try this
Write down the maximum earth fault loop impedance for the following:
1. For circuits containing socket outlets
a. BS88 Part 2 and Part 6
    i) 10 A

    ii) 20 A

    iii) 32 A

b. Type 2 MCB to BS 3871
    i) 5 A

    ii) 15 A

    iii) 30 A

2. For circuits supplying fixed equipment only:

a. BS88 Part 2 and Part 6
    i) 10 A

    ii) 20 A

    iii) 32 A

b. BS 1361
    i) 5 A

    ii) 15 A

    iii) 30 A

If the circuit being tested is protected by a residual current device there may be a problem with the device tripping out each time a test is carried out. There are instruments that allow the test to be made without the device operating.

Where such an instrument is not used the earth fault loop impedance is derived by

- measuring the earth fault loop impedance on the supply side (upstream) of the RCD
- measuring the $R_1 + R_2$ of the circuit downstream of the device, at the furthest point on the circuit from the supply
- adding the two values together to determine the overall Z_s of the circuit.

When we determine the overall Z_s of a circuit in this way we need to make a note on the certification to inform the recipient that the value was measured in this way.

Remember

The results of an earth fault loop impedance test should be checked against the maximum values in the tables.

Testing for earth fault loop impedance can be made from any point on the circuit but the value which is recorded is the highest value on the circuit. This is the worst case scenario for the circuit where the loop impedance is at its highest and the fault current will be at its lowest. Such conditions are generally found at the point on the circuit furthest from the supply.

Testing earth fault loop impedance on BS 1363 socket outlet circuits is quite straight forward as a lead is provided with the instrument, complete with plug. This is simply inserted in the socket in order for the test to be carried out. The instrument carries out some initial checks

- is the polarity of the connection correct and
- is a suitable connection to earth available?

If the circuit under test fails either of these checks then the necessary remedial work must be carried out before testing can begin. When the instrument indicates that all is well the test can proceed and the reading taken.

The instrument injects a current of about 25 A through the circuit under test, from the earth connection around the fault path shown in Figure 8.27, and back via the phase connection. Whilst this test is being carried out the potential on the exposed and extraneous conductive parts throughout the installation will be at a potential above earth potential. If the circuit is incomplete and the test is undertaken then there is a real risk of electric shock if someone inadvertently becomes part of that circuit. Equally if a person forms a connection between the installation earthing system and exposed conductive parts and true earth there is also the risk of electric shock. It is important that we ensure the circuit is complete and warning notices are placed to advise anyone in the vicinity that testing is being carried out and that the installation should not be used.

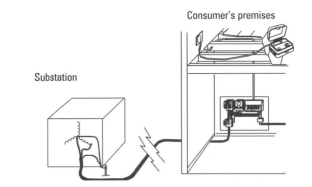

Figure 8.23   *An earth fault loop impedance test carried out on an installation. The substation supplies the installation via an underground cable.*

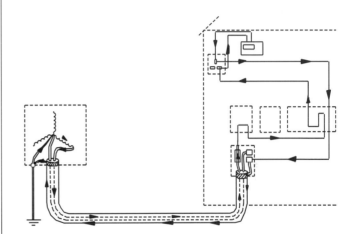

Figure 8.24

Testing circuits which supply equipment other than socket outlets follows the same procedure but requires the use of a set of proprietary test leads, as shown in Figure 8.25, to make the connection to the circuit.

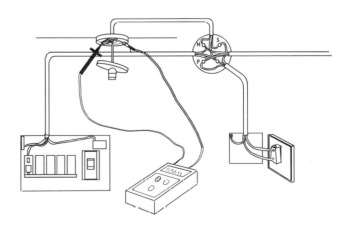

Figure 8.25   *A set of proprietary test leads connected to a suitable light fitting*

Again the value we record should be the highest obtained on each circuit.

The test can be taken from any supply point and should be carried out on all types of circuit. A socket outlet circuit is straight forward as the test equipment just has to be plugged in.

Once we have recorded the value of earth fault loop impedance we need to check these against the maximum acceptable test values for the type of protective device used. The values we considered earlier in Tables 41 from BS 7671 are the design values for the circuit and these values are calculated for when the circuit is operating under normal conditions, ie carrying load current. When we carry out our test the circuit is not loaded and therefore the conductor is operating at a much lower temperature than it would be on load. Also the fault current which flows will produce a further rise in temperature which our test will not be able to simulate. For these reasons we cannot use the maximum values of Z_s given in the tables. As a rule of thumb the maximum values of Z_s for test results should not exceed 75% of the values given in BS 7671.

For example if the protective device is a BSEN 60898 32A type B circuit breaker the maximum design Z_s from Table 41B2 is 1.5Ω. The maximum value acceptable as a test figure is going to be 75% of 1.5 = 1.5 × 75 ÷ 100 = 1.125 Ω.

When we check the results of our test they should not exceed the modified maximum test value, calculated by the rule of thumb above. Should the values exceed the maximum then further investigation is necessary to establish the cause and remedy the situation before the installation can be put into service.

Polarity checks

Having tested for polarity prior to the supply being connected we need to test the installation to verify that the polarity is in fact correct. This may be confirmed whilst carrying out the earth fault loop impedance tests.

Earth electrode resistance test

Every installation that forms part of a TT system will have its own earth electrode forming part of the earth fault return path to the supply companies transformer. We need to test the resistance of this electrode to the general mass of earth to ensure that the installation meets the requirements of BS 7671. There are a number of ways to carry out this test and to some extent it will depend upon the location of the installation as to which is the most appropriate test to use. We shall consider the two most common methods of testing the earth electrode the first being the use of an earth fault loop impedance test instrument, the second being a proprietary earth electrode test instrument.

For safety, as the test calls for the electrode to be disconnected from the main earth terminal, both methods require the installation to be isolated from the supply before the test procedure is started. It is preferable to disconnect the electrode from the earthing conductor, but where this is not possible the earthing conductor may be disconnected from the main earthing terminal.

Earth fault loop impedance test instrument method

As this method requires the use of the incoming supply to provide the power for the instrument it may prove to be advantageous to isolate the installation from the supply and disconnect the earthing conductor from the main earthing terminal and connecting the instrument as shown in Figure 8.26.

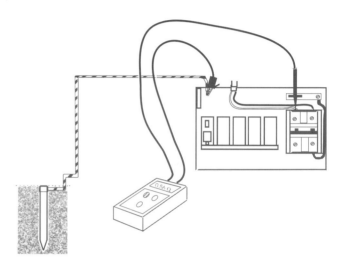

Figure 8.26

Once the earth electrode is isolated from the installation main earthing terminal the earth electrode resistance can be measured in the same way as we measured Z_e for the TN installations. In this case, however, as we are testing the resistance of the electrode to the earth and the return path through the mass of earth to the supply transformer electrode and back, via the phase conductors to the installation, the reading on the instrument is likely to be much higher than that recorded on the TN systems.

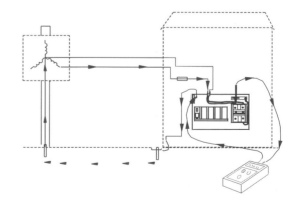

Figure 8.27   *Testing to the earth electrode measuring the earth fault path which gives an indication of the earth electrode resistance.*
To simplify the drawing the consumer and supply company's main equipment has been omitted.

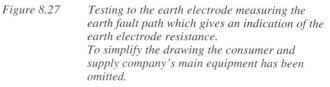

Proprietary earth electrode test instrument method

This method involves the use of a proprietary test instrument, specifically designed for the purpose. There are two main types of instrument for this one having three terminals the other having four terminals. The four terminal type requires two of the terminals to be linked together and you should refer to the particular manufacturer's instructions for this. The remainder of the requirements are the same for both tests and so we shall consider the test instrument with three terminals for ease of reference.

Remember
The termination references for the instrument may vary from one manufacturer to another so always check the instructions to ensure the connections are correctly made before testing begins.

For this test it is advantageous to make the connection directly to the earth electrode and where possible the earthing conductor should be disconnected from the electrode once the installation has been isolated from the supply. If this is not possible the earthing conductor should be disconnected from the main earthing terminal and isolated. The test should then be carried out at the earth electrode.

A second electrode is driven into the ground some 20–30 m from the installation electrode which is under test. A third electrode is then driven in halfway between the two electrodes as shown in Figure 8.28.

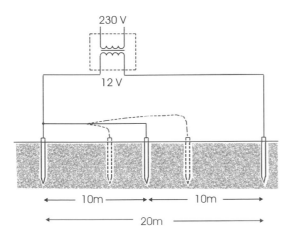

Figure 8.28   Measurements may be adjusted to suit the location

The three terminal instrument is then connected with the terminal marked "E" to the installation electrode, the terminal marked "C" to the electrode furthest from the installation electrode and the terminal marked "P" to the central electrode, as shown in Figure 8.29 using separate leads for each electrode.

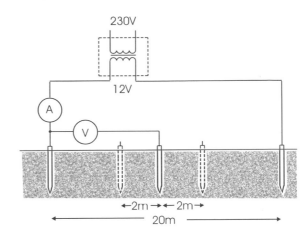

Figure 8.29

The principal of the test is that the instrument passes current between the two outside electrodes and measures the voltage drop across the connection to earth of the installation electrode and the centre electrode. This provides enough information for the instrument to measure and record the resistance of the electrode, in ohms, on the instrument scale. This test is repeated with the centre electrode approximately 3 m closer to and 3 m further away from the installation electrode. If these three tests produce results that are substantially the same, say within about 5%, then average of the results should be calculated and recorded as the electrode resistance for the installation.

If the results are not substantially in agreement it indicates that the resistance areas of the electrodes are overlapping. The furthest electrode from the installation electrode must be moved further away and the tests repeated as before.

The main drawback to this method of testing is that it requires a considerable amount of open space adjacent to the installation. This space needs to be suitable for the installation of the test electrodes at the three locations, so metalled car parks, footpaths and the like are not suitable. For this reason the locations at which this method of testing electrode resistance are relatively few and the test is generally carried out using an earth fault loop impedance test instrument.

Testing residual current devices

There are many occasions where RCDs are used to provide additional protection against electric shock from direct or indirect contact. Wherever such a device is installed we must test to ensure that disconnection is achieved within the time required to provide protection from electric shock.

As we established earlier, the two most common ratings for RCDs are 100 mA and 30 mA, and both types require testing to ensure correct operation. There are two tests which need to be carried out on each device, the operation when the periodic test button is operated and the simulation of a fault by using an RCD test instrument. It is important that the electrical tests are carried out before the test button is checked as this may effect the performance of the device.

The RCD test instrument needs to supply a range of test currents appropriate to the device being tested and display the time taken for the device to operate. One of the principal requirements is that under any circumstances the duration of the test current cannot exceed 2 seconds, this makes it possible to test most time delay devices which have been installed to ensure discrimination between RCDs of different operating currents. In addition most RCD test instruments have the facility to carry out the test in either the positive or negative half cycles of the supply. Each of the tripping tests should be carried out in both half cycles and the highest value obtained should be recorded.

Remember

The test instrument will need to have leads suitable to carry out the tests on circuits supplying both socket outlets and fixed equipment. We shall need leads with both a moulded plug and split leads, similar to those used for the earth fault loop impedance tester.

There are three basic tests which should be carried out in this instance we are going to consider the testing of a device rated at 30 mA.

Half rating test

This test is intended to establish whether the device is likely to suffer from nuisance tripping under normal operation. The RCD test instrument is set to half the rated tripping current for the device under test, in our example this would be 15 mA, and the instrument is connected as shown in Figure 8.30.

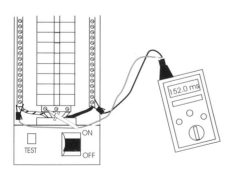

Figures 8.30

The test is then carried out and the result noted. For this test the instrument should apply the test current for the full 2 seconds without the RCD operating, if the device should trip it is possible that there could be nuisance tripping and the manufacturer of device should be consulted to establish the operating parameters of the device and whether a replacement should be fitted.

The instrument is then set to the full operating current of the device, in our case 30 mA, and the test applied. This time the device should operate and the time taken displayed on the instrument, this should be noted and the device reset. The instrument should then be set to test on the other half cycle and the test repeated and again the result should be noted. The highest of the two results is the one which should be recorded on the Schedule of Test Results.

We need to check the recorded values against the maximum operating times for the type of device installed. As the criteria for acceptable operating times will depend on the product standard we shall need to refer to the maximum for the particular standard. Table 8.1 gives the maximum values for the product type when tested at the full rated current for some of the most common RCDs.

Table 8.1

RCDs to BS 4293 RCD protected sockets to BS 7288	Less than 200 ms
RCDs to BS 4293 (with a time delay incorporated)	Between : 200 ms + 50% of the time delay and 200 ms + 100% of the time delay
RCDs to BS EN 61008	Less than 300 ms

The third test to be carried out is applicable to RCDs used to provide protection against direct contact and rated at no more than 30 mA. For these devices we must carry out a test at five times the rated current, so in the case of our 30 mA device this will be at 150 mA. When this test is applied the device should operate within a time of 40 ms, that is 0.04 s.

Once the above tests have been completed the device should be reset and the test button operated to ensure the device operates.

Remember

The RCD test button is the monthly check that should be undertaken by the user of the installation and simply checks the device operation it does not confirm satisfactory disconnection times for protection against electric shock.

Functional tests

We are expected to carry out functional tests of equipment such as switches, interlocks, drives and controls to ensure they operate correctly. This will include checking the operation, mounting, accessibility and adjustment of equipment and devices. There is no formal record of each item required however there is a need to indicate that the functional testing of equipment has been undertaken.

Certification

Throughout the inspection and testing activities reference has been made to the recording of test results. These should be detailed on the schedule of test results and be included with the certification provided for the installation. There are three principal forms of certification and each has a specific purpose and so care must be taken to ensure the correct documents are issued. As the appropriate form of certification must be issued to the person requesting the work we shall briefly consider the forms and the purpose of each.

Electrical Installation Certificate

We are required to issue an Electrical Installation Certificate, including a schedule of test results, for all new installation work, including alterations and additions to an existing installation. A separate certificate must be issued for each separate electrical installation, so a block of six flats, for example, would require seven completion certificates, one for each flat and one for the landlords areas such as stairs and hallways. The Electrical Installation Certificate must be issued to the person who requested the work, who may not be the final user of the installation, whether the certificate has been requested or not.

The Electrical Installation Certificate provides details of the persons responsible for the design, construction and inspection and testing of the installation. The individuals responsible for each aspect of the installation are required to sign the relevant section to confirm compliance with the requirements of BS 7671. The Electrical Installation Certificate should be issued by the contractor responsible for the construction of the installation and those responsible for the design and inspection and testing aspects should complete the relevant sections.

Where the electrical contractor is responsible for all three aspects of the installation there is a short form of the certificate which requires a single signature covering the design, construction and inspection and testing of the completed installation.

Minor Electrical Installation Works Certificate

The Minor Electrical Installation Works Certificate is for use when the work undertaken

- is related to a single circuit only
- does not involve the installation of a new circuit or protective device

and typically would be used for additional sockets or lights to an existing circuit. The certificate contains the relevant details for the circuit involved and a section for the result of the essential tests that need to be carried out on completion of the work.

The use of the Minor Electrical Installation Works Certificate needs to carefully considered as a certificate is required for each circuit which is worked on. If the alteration or additional work involves more than one circuit it may be more beneficial to issue an Electrical Installation Certificate which clearly identifies the extent of the work carried out.

Remember that the replacement of a consumer unit or distribution board requires a Electrical Installation certificate as each circuit will have been worked on and it is usual that the protective devices are changed and may be of completely different type.

Periodic Inspection Report

The Periodic Inspection Report is used to report on the condition of an existing electrical installation which has been energised and put into service. There are a number of reasons why such a report may be required, such as a house purchase, entertainments licence, insurance or to establish the condition of the installation and whether any remedial or improvement work is required. All installations should be inspected and tested at regular periods to ensure that the installation has not been damaged or deteriorated over time. The time between these inspections is dependent upon a number of factors, such as the type of installation and its use, the environmental conditions which exit and any special requirements such as frequent change of occupancy.

A Periodic Inspection Report is not an appropriate document to issue for a new installation and does not provide the declaration in respect to the design and construction of the installation or that the installation was inspected and tested during its construction. The Periodic Inspection Report does not confirm that the installation is safe to use, what it should provide is a report on the compliance of the electrical installation with the current requirements of BS 7671. The report should highlight any departures from the requirements and the urgency of any remedial work. The report should detail in layman's terms where the installation does not comply and not what remedial action is needed to put it right.

The electrician carrying out the report should make an overall observation on the general condition of the electrical installation and whether the compliance of the installation with the requirements of BS 7671 is satisfactory or unsatisfactory. Once again the report should be issued to the person requesting the work and it is not necessary for remedial work to be completed before the report is issued. Any remedial work carried out should be certificated on either an Electrical Installation or Minor Electrical Installation Work Certificate as appropriate.

Remember

- All electrical installation work should be certified using the appropriate form prior to it being placed in service.
- Any inspection and test on an existing installation which has been in service should be certificated on a Periodic Inspection Report.

Points to remember ◄ ---------------

The tests are all related to the safety and operation of the circuit equipment.

We test to ensure that all fuses, switches and other devices that may break the circuit are all connected into the phase conductors.

When a fault develops between phase and earth we expect the protection device to operate. The _____ _____ _____ _____ test is carried out to ensure that this can happen.

Where an earth electrode is used as a means of consumer's earth return it must be tested to ensure it is of a low enough _____ to clear any possible faults.

Where residual current devices are used to protect against _____ _____ _____ they must be tested to confirm they will still operate within their range under fault conditions.

A functional test shows whether assemblies, such as switch gear and control gear, are _____ _____.

All electrical work must be suitably certificated.

Self-assessment short answer questions

All questions in this section refer to the specification and plans of the small factory shown in the Appendix.

1. Explain how a polarity test can be carried out to the lights in the reception area.

2 Explain how an earth fault loop impedance test can be carried out in the first floor offices from
    a. a 13 A socket outlet
    b. a fluorescent fitting

3. An outbuilding is to be erected in the south corner of the car park. As other services are to be installed in the building it is proposed to make the electrical installation for the outbuilding TT. Explain two methods of carrying out an earth electrode test and list some of the problems which may be encountered, particularly as access to adjacent properties cannot be obtained.

4. The socket outlet on the loading bay is protected by a 30 mA RCD. Detail the tests required to confirm the operation of this device in order to provide protection against electric shock and list the maximum acceptable values.

Part 5

Equipment testing

So far we have considered the testing of the electrical installation fixed wiring and it is important to ensure that the fixed wiring meets the required standard. However there are parts of the installation which are not part of the fixed wiring yet need to be tested, that is the fixed equipment associated with the fixed wiring.

The person responsible for an electrical installation for property used by other people, known as the "Duty Holder", has an obligation to ensure that any portable or transportable appliances are kept in good condition and safe to use. This equipment is not associated with the fixed wiring and is subject to more frequent change than equipment connected directly to the fixed installation. The considerations for the inspection and testing of this so called "portable and transportable equipment" is covered in the Code of Practice for In-service Inspection and Testing of Electrical Equipment, published by the IEE.

We shall be looking at the requirements for fixed and transportable equipment in this lesson and it would be a good idea to have a copy of the code of practice to hand when working through the requirements.

Fixed equipment

Most installations have some fixed equipment connected to them and during the testing of the installation the equipment must be isolated or disconnected. But before previously used equipment is put back to work it must also be inspected and tested.

Examples of fixed electrical equipment in a domestic installation would include: the cooker, immersion heater, shower unit, storage radiator and so on.

Where the equipment is clearly Class 1 equipment, in other words it has an exposed metal enclosure and does not have the double insulated Class 2 symbol, then an insulation resistance test must be carried out.

A 500 V d.c. insulation resistance tester can be used for this, connected between the live conductors and the metal case. If there is a British Standards number on the equipment then that standard will give the minimum value of insulation resistance acceptable. If there is no British Standards referred to then the minimum acceptable resistance is 0.5 MΩ.

Where equipment is disconnected from the installation and is tested for insulation resistance the resistance should not be less than 0.5 MΩ. The test is taken between the supply conductors and the circuit protective conductor and frame of the equipment.

Figure 8.31   *An insulation resistance test carried out on a three-phase motor.*

Manufacturers' instructions need to be followed closely if the equipment being tested contains any electronic components. A 500 V insulation resistance test can destroy devices that are not designed for that test voltage.

Portable electrical equipment

The Health and Safety Executive found that about 25% of all reportable electrical accidents involved portable equipment. The majority of these accidents caused electric shock but many others resulted in burns from arcing or fire. Typical accidents were caused by:

- equipment being used for jobs for which it was not designed
- inadequate maintenance or misuse
- the use of defective apparatus

In an attempt to overcome this a number of recommendations were produced and this ultimately resulted in the production of the Code of Practice for in Service Inspection and Testing of Electrical Equipment. This document provides guidance on the procedures for regular inspection of equipment and the need and frequency for testing to be undertaken. The Code also provides details on the methods of testing and acceptable results, in the absence of any manufacturers requirements.

The Code recognised that the frequency of inspection needs to be more regular than the formal testing of equipment. It also acknowledges the there are checks which need to be carried out by the user of the equipment. The first requirement is to set up a register of equipment and record the date that each inspection and each inspection and test is carried out.

As with the requirement for periodic inspection the periodicity of the inspection and the testing is dependent on the type of equipment and the type of premises in which it is used. Table 1 in the Code of Practice for in Service Inspection and Testing of Electrical Equipment provides suggested periods for the initial inspection and test of types of equipment and their environments.

The inspection of portable equipment should include:

- looking for signs of damage or deterioration
- examining the casing
- checking the plug for
  - case damage
  - damaged pins
  - faulty or loose terminals
  - cable anchoring device
  - heat discolouration
- examining the cable sheath
- checking terminals
- inspecting any control gear used in association with it
- where applicable inspecting brush gear and commutators

After the visual inspection has been carried out and it is considered that the equipment is safe it may then be tested. The tests fall into two categories - those that are essential and those that are optional. All of the tests can be carried out using separate instruments and some of the instruments used to test the fixed wiring may also be suitable for some of these tests. There are however some tests that may require special equipment designed specifically for the purpose. This has led to instrument manufacturers developing special Portable Appliance Test (PAT) equipment which is capable of carrying out all of the required tests. The complexity of this equipment varies between manufacturers but generally the tests are similar.

Some of these instruments have the facility to record the data on an integral processor unit and these can later be downloaded to produce a printed report and record for each item of equipment. Instruments are also available that print identification labels for the equipment identifying when the test was carried out and providing a reference number, often in the form of a bar code.

The record of each individual item and the details required may be recorded on record sheets, proformas of which are given in the Code of Practice for in Service Inspection and Testing of Electrical Equipment. Other forms may be used, but once again these should record at least all the information contained on the proforma documents in the Code of Practice.

The tests

For the purpose of this exercise we shall consider the testing of equipment using a proprietary Appliance Tester. It is possible to carry out the tests using individual instruments, their use is covered in the Code of Practice, and is similar to that used for the fixed wiring tests and so we shall not consider that method here.

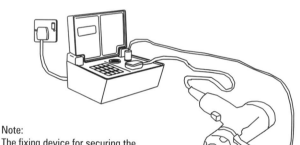

Note:
The fixing device for securing the drill has been omitted for clarity.

Figure 8.32   *The metal cased electric drill should be securely fixed before the earth bond test is carried out.*

Earth continuity test

This test can only be carried out on Class 1 equipment. This means that the frame is made of conducting material which is used as a protective conductor. The test is designed to ensure that any exposed metalwork on the equipment under test is effectively connected to a safe earth potential. A high current of about 25 A is passed through the test circuit at about 6 V while the resistance is measured.

The equipment under test is plugged into the test instrument and the test instrument is then plugged into a mains supply. A test lead is connected from the test earth bond terminal to the exposed metalwork of the equipment under test.

When the test button is pressed the high current flows from the test instrument through the earth bond lead to the metal frame, through this to the connections of the protective conductor and then back to the test instrument. If all of the connections are satisfactory the resistance should be between 0 Ω and 0.1 Ω. Where low current equipment with thin flexible cables are tested the reading may be up to 0.5 Ω. In general terms equipment that has a power rating of less than 1 kW, has a thin mains lead and is protected by a 3 A or 5 A fuse, would fit into the category where the resistance could be up to 0.5 Ω. In other cases the resistance should not be greater than 0.1 Ω. Where equipment has long leads connected to it there may be problems in getting down to 0.1 Ω. In these cases calculations may have to be carried out to see if the protection device on the equipment would operate in the event of a fault developing.

Table 2 of the Code of Practice for in Service Inspection and Testing of Electrical Equipment provides details of the acceptable values for the earth continuity for appliances and equipment dependent upon type and standard to which they are manufactured.

Insulation resistance

Where double insulated Class 2 equipment is being tested this may be the first test depending on the manufacturer of the test equipment. As Class 2 equipment does not have exposed metal some test equipment does not include the insulation resistance test for Class 2 equipment. However we will look at how this test can be carried out on both Class 1 and Class 2 equipment.

The test voltage is applied at 500 V d.c. and the resistance should not be less than 2 MΩ. The equipment must be switched ON when this test is applied.

Class 1 equipment

If the equipment failed Test 1 then this test should not be applied.

On earthed equipment the test is carried out between the earth pin and the combined live and neutral pins of the plug.

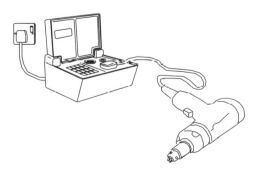

Figure 8.33

When the test button has been pressed the resistance reading should be noted. It is not sufficient to say if the equipment has passed or failed.

Class 2 equipment

As double insulated equipment does not have a connection to the case some test instruments have a probe that can be held against it during the test. This ensures that no part of the case is conductive.

Figure 8.34

Several tests may need to be carried out with the probe held in different positions.

Table 3 of the Code of Practice for in Service Inspection and Testing of Electrical Equipment gives details of the acceptable values of insulation resistance for appliances and equipment dependent on type and standard to which they are manufactured.

On some Appliance Test Instruments there is a facility to carry out earth leakage testing of equipment, and this may be substituted for insulation resistance testing when that test is inappropriate. This test involves the measurement of current as opposed to the measurement of resistance. The test instrument monitors the earth leakage while the equipment is operating. This is achieved by either measuring the current flowing in the earth conductor, or by comparing the current flowing in the live to that in the neutral, depending on the test instrument used.

Dialectric strength testing

This is also known as the "flash test" and is not normally carried out during the in service testing.

Functional testing

This, as the name implies, is intended to establish whether the equipment is working properly. Without the availability of sophisticated equipment the functional test is an acceptable method of determining the correct operation of the equipment. Load testing may be useful to determine whether items, such as a unit containing a number of heating elements, has one or more of its elements open circuit.

The test is carried out at reduced voltage to check that the power consumption is similar to that shown on the equipment rating plate. The equipment must be switched on throughout the test. The test is the same for both Class 1 and Class 2 equipment and consists of a low voltage, about 6 V a.c., being applied across the L and N of the mains plug. The test instrument usually gives a reading in watts that can be compared with the rating plate.

When carrying out testing on in service equipment it is important to ensure the tests carried out are appropriate and that the results are acceptable. The Code of Practice for in Service Inspection and Testing of Electrical Equipment provides guidance on the type of equipment, the appropriate tests and the acceptable results. In addition it provides details of the responsibilities of those involved in all aspects of the maintenance and checking of equipment and the requirements of the law.

If you are to be involved in this particular activity it is advisable to obtain a copy of the Code of Practice for in Service Inspection and Testing of Electrical Equipment, familiarise yourself with the requirements and ensure that the reports you issue are in accordance with the model forms in the Code.

In addition to the requirements for in service inspection and testing of equipment and appliances many items of equipment include electronic components which may require specialist testing and equipment. In this next lesson we shall consider the testing of such equipment, some of the specialist equipment required and the precautions we need to take.

Electronic equipment

It is becoming more and more important to be able to recognise how to carry out some tests on electronic equipment. The equipment could be a comparatively simple regulated power supply unit or a more complex programmable control circuit. The important thing is to be able to carry out tests using the correct test equipment without damaging either the equipment under test or the test equipment.

Testing a regulated d.c. supply unit

The regulated supply unit shown in the circuit diagram, Figure 8.35, is supplied with 230 V a.c. and has a d.c. output which should be 15V d.c. either side of zero.

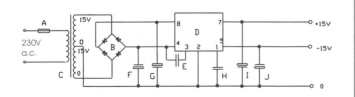

Figure 8.35

As the circuit diagram bears very little relationship to the actual layout of the components as shown in Figure 8.36 it is important to be able to relate one to the other.

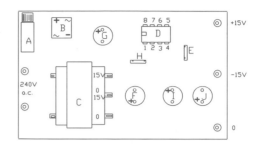

Figure 8.36

When making tests on a circuit like this it is important to have some idea what the answer should be. This helps to identify the instrument to be used and the range it should be set to.

First let's assume we want to take four sets of readings to
• check the supply voltage
• check the output from the transformer
• check the output from the rectifier
• measure the output voltages from the unit

Having decided what to measure a suitable instrument needs to be selected. As there are several different voltages, both a.c. and d.c., required on these tests the most suitable instrument is a multi-range type. This must be able to measure at least 240 V a.c. and give accurate readings down to 15 V a.c. as well as measuring readings of 15 V d.c. There are many instruments capable of giving these ranges, some using the traditional analogue scale others giving a digital read out. In many cases such as the example we are looking at here it does not matter which type of instrument is used. There are however applications where only the digital type would be suitable. These are where the current drawn by an analogue instrument is too great for the components being tested. The moving coil movement of an analogue instrument draws a current to give the needle deflection. This current, when drawn through components designed for micro-amps, can destroy the properties of the component. In these circuits electronic digital type instruments must be used.

Digital multimeter

Analogue multimeter

Cathode ray oscilloscope

Battery charger

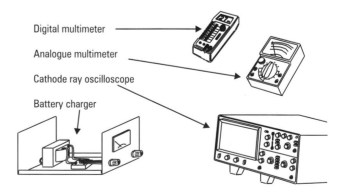

Figure 8.37   *Instruments that can be used to measure voltages*

When taking voltage readings of this type safety precautions must be taken.

- The correct type of test probes must be used.
- Care must be taken so that when the instrument is to test one point it is not possible to touch other exposed connections.
- Insulated mats may need to be laid over bare terminals so that they cannot be touched inadvertently.

Taking the readings

As the readings we are taking are voltages there is no need to disconnect any components. The test instrument needs to be switched to a.c. at a voltage greater than 240 V. The supply voltage can be measured across the supply connections or across the input to the transformer. The second location will also check if the fuse is satisfactory.

The second readings are also on a.c. but before they are taken the 230 V terminals should be covered so that they cannot be touched. The output of the transformer is 30 V centre tapped so as to give two 15 V a.c. supplies. Some electronic digital instruments are self ranging and need only to be set to the a.c. position. On others the appropriate scale must be selected.

The third reading is the output of the rectifier so it is a d.c. measurement and the instrument needs to be switched accordingly.

The fourth reading is also d.c. but this time two readings have to be taken and it is important to ensure the polarity of the test instrument is correct for each.

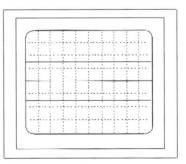

When testing electronic circuits dangerous voltages may be present.

Take precautions to avoid contact with any exposed metalwork that may pose a danger.

The previous voltage readings will give an indication as to the output voltage of the circuit but they will not show how effective the rectifier or the smoothing circuit is. To know how good they are we need to be able to see a picture of what is happening. This is possible with an oscilloscope which displays a trace on a cathode ray tube. Now let us look at the outputs on an oscilloscope screen for each of the following.

- input to the rectifier
- output from the rectifier
- output voltages from the unit

The input to the rectifier is 30 V a.c. (r.m.s.). This means that on the oscilloscope we should have an a.c. waveform showing 50 Hz with a peak to peak voltage of $1.414 \times 30 \times 2$.

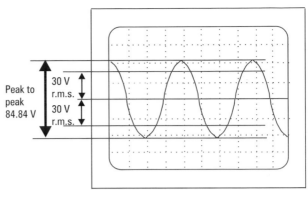

Figure 8.38   The input to the rectifier

The output of the rectifier shows a full wave rectified output with the bottom half of the waveform made positive.

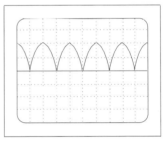

Figure 8.39   The output from the rectifier

Looking at the d.c. output of the unit each 15 V is either side of zero. Using a two beam oscilloscope both of the 15 V can be displayed at the same time. Each trace is a straight line indicating that all of the a.c. content has been smoothed out.

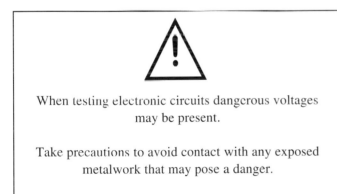

Figure 8.40   The output voltages from the unit

On some circuits, such as amplifiers, the input has to be applied at a high frequency. This is usually obtained through a signal generator which can be adjusted to give the frequency required. In this case the signal generator is connected to the input of the amplifier and an oscilloscope measures the output. From this the amount of amplification the circuit can give can be checked.

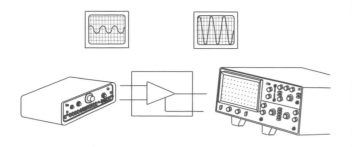

Figure 8.41   *The amplitude of the input signal is increased within the amplifier, producing an output of the same frequency but with a higher amplitude.*

Frequency can also be measured using an oscilloscope measuring the cycles and calculating it from the time base used, but a frequency may be more suitable. Frequency meters can typically measure ranges from 10 Hz to 20 MHz and the frequency can be read direct from a digital display.

Logic probes

So far we have used an oscilloscope to view how a voltage within a circuit varies with time – the voltage varies continuously and is shown as a waveform (Figure 8.42).

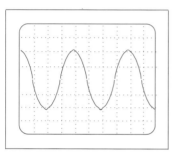

Figure 8.42

These signals are known as **analogue** signals.

Many electronic devices use **digital** signals. A digital signal is one that can only be in one of two states, either a high voltage or a low voltage. On an oscilloscope these signals would be shown as in Figures 8.43.

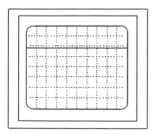

A "high" voltage signal

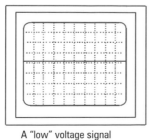

A "low" voltage signal

Figure 8.43

In a high speed digital electronic circuit these signals will change very rapidly between high and low. On an oscilloscope the trace would appear as shown in Figure 8.44.

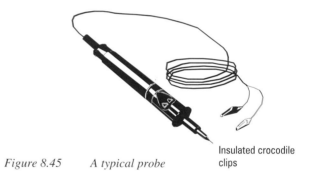

Figure 8.44

Although a digital circuit can be tested using an oscilloscope there is a much simpler and cheaper tool available. A **logic probe** is small, compact and very much cheaper than an oscilloscope. Figure 8.45 shows a typical probe which has an exposed tip for placing on the circuit to be tested.

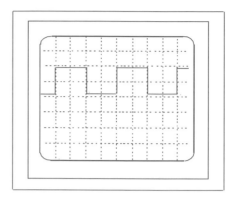

Figure 8.45   *A typical probe*   Insulated crocodile clips

The logic probe will have 2 LEDs (often red and green) which indicate the logic states high and low. If the signal is changing rapidly between high and low both LEDs will light but on half brightness. On some logic probes there may also be another LED which will indicate that the signal is changing.

If the signal is floating (NO LEDs) then no component is driving it high or low.

There are two types of digital signals that the logic probe may be used with – TTL and CMOS. You should use the correct logic probe for the type of circuit. In some cases the logic probe will have a switch to select TTL or CMOS. If in doubt use the TTL setting.

Points to remember ◀ – – – – – – – – – –

The Electricity at Work Regulations have made it a legal requirement for equipment to be regularly inspected and tested.

The tests that meet the legal requirements are basically insulation resistance tests but instrument manufacturers have made it possible to carry out further tests that can verify the working of the equipment as well as the safety. In addition to the legal requirements other tests may need to be carried out on equipment by electricians.

List the tests covered in this chapter:

Self-assessment short answer questions

1. The following equipment has to be tested. State what instrument, and its range, would be most suitable for each.
   a. the voltage output of a battery charger
   b. the current flowing when a battery is on charge
   c. the input and output signal of an audio amplifier

2. Draw on the oscilloscope screen below an a.c. waveform with an r.m.s. voltage of 25 V and a frequency of 50 Hz.

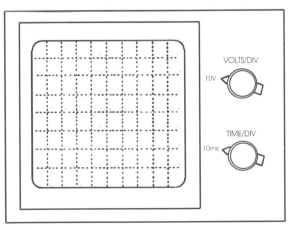

Self assessment multi-choice questions

1. A continuity test is most likely to use an instrument with a range of
   a. $0 - 10\ \Omega$
   b. $0 - 1\ k\Omega$
   c. $0 - 1\ M\Omega$
   d. $0 - 100\ M\Omega$

2. A ring final circuit wired in 2.5 mm² cable makes up a total length of 40 m. When measuring the continuity of the phase conductor the resistance will be approximately
   a. $0.05\ \Omega$
   b. $0.5\ \Omega$
   c. $5.0\ \Omega$
   d. $50\ \Omega$

3. The wording on the label of a supplementary bonding conductor reads
   a. Warning – do not disconnect
   b. Electrical earth – do not remove
   c. Earth bonding conductor – do not disconnect
   d. Safety electrical connection – do not remove

4. The main bonding conductors should be connected to the gas and water pipes at a point
   a. where the pipes enter the building
   b. on the supply side of the stop valves
   c. on the consumer's side of the stop valves
   d. somewhere along their length

5. The insulation resistance test voltage on a 240 V circuit should be at least
   a. 250 V
   b. 500 V
   c. 750 V
   d. 1000 V

6. A reading on an insulation resistance test instrument that shows ∞ or INF means that
   a. the instrument is faulty
   b. the batteries in the instrument require replacing
   c. the resistance is higher than the range can indicate
   d. the resistance is lower than the range can indicate

7. When an insulation resistance test is carried out on equipment disconnected from the installation the minimum acceptable resistance between phase and earth is
   a. $0.5\ \Omega$
   b. $0.5\ k\Omega$
   c. $0.5\ M\Omega$
   d. $5.0\ M\Omega$

8. When a probe 12 mm in diameter and 80 mm long cannot penetrate an enclosure, the enclosure is said to conform to
   a. IP2X
   b. IP4X
   c. IP6X
   d. IP8X

9. All switches and fuses on a three-phase circuit should be connected in
   a. the neutral only
   b. the non-earthed conductor
   c. the earthed conductor
   d. any two phases

10. In a.c. circuits switches can be connected into the neutral conductor
    a. at any time
    b. providing the neutral is fused
    c. only if mechanically linked to a switch in the phase conductor
    d. only if mechanically linked to a switch in the cpc.

11. An earth fault loop impedance test verifies the impedance of the path in the event of a fault between
    a. phase to neutral
    b. neutral to cpc
    c. cpc to earth
    d. earth to phase

12. The current injected into a circuit when an earth fault loop impedance test is carried out is subject to a maximum of
    a. 100 A
    b. 50 A
    c. 25 A
    d. 5 A

13. (1) Pressing the test button on an RCD is the same as carrying out the recommended tests.
    (2) A 30 mA RCD should trip at a 15 mA leakage to earth
    a. only statement (1) is correct
    b. only statement (2) is correct
    c. both statements are correct
    d. neither statement is correct

14. The first test that should be carried out on Class 1 portable appliances is
    a. insulation resistance
    b. flash test
    c. operational test
    d. earth bond test

15. The test voltage for an insulation resistance test on portable appliances is
    a. 6 V
    b. 200 V
    c. 500 V
    d. 2000 V

16. An oscilloscope connected to the output of a full wave rectifier will look like
    a.

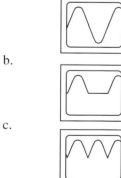

    b.

    c.

    d.

17. A voltage described as 230 V a.c. r.m.s. will have a peak to peak value of
    a. 230 V
    b. 325.32 V
    c. 460 V
    d. 650.44 V

18. A logic probe is a diagnostic tool which can be used to indicate the
    a. current flowing in a circuit
    b. power dissipated in an I.C.
    c. the state of a digital signal
    d. temperature of an I.C.

Answer grid

1	a	b	c	d		10	a	b	c	d
2	a	b	c	d		11	a	b	c	d
3	a	b	c	d		12	a	b	c	d
4	a	b	c	d		13	a	b	c	d
5	a	b	c	d		14	a	b	c	d
6	a	b	c	d		15	a	b	c	d
7	a	b	c	d		16	a	b	c	d
8	a	b	c	d		17	a	b	c	d
9	a	b	c	d		18	a	b	c	d

All these questions refer to the specification and plans of the small factory shown in the Appendix.

1. The ring final circuit in the first floor office is supplied from the consumer unit on the first floor landing. Explain how a continuity test for this ring final circuit could be carried out.

2. In the boiler room there are a great many pipes. Explain how a continuity test can be carried out to ensure that all of the pipes are at the same potential.

3. The steel conduit through to the light in the works office has to be tested to check it is suitable as a circuit protective conductor. List the special requirements the test equipment must have to be suitable for the test.

4. In the kitchen there is a great deal of stainless steel exposed to touch. Explain how a continuity test should be carried out to ensure that this is all at the same potential as the "electrical earth".

6. The lighting in the reception area has just been rewired. Describe, with the aid of a sketch, how insulation resistance tests should be carried out before the supply is reconnected.

5. An insulation resistance test is to be carried out on the main factory lighting circuits. Explain what preparations must take place before the tests can be carried out.

7. A bench in the factory has electronic equipment permanently wired into switched fused spur plates. Explain how an insulation test can be carried out on these circuits without damaging the electronic equipment.

8. A visual inspection reveals that an MICC cable in the boiler house has been dented. Explain how you would test this cable to ensure that it is safe to continue using.

9. Explain how a polarity test can be carried out to the lights in the reception area.

10. Explain how an earth fault loop impedance test can be carried out in the first floor offices from a
    a. 13 A socket outlet
    b. fluorescent fitting

11. A lightning protection electrode is installed outside the building as shown on the plan. Explain some of the problems that must be considered before an earth electrode test can be carried out.

12. The socket outlet on the loading bay is controlled by a 30 mA RCD. Explain how this should be tested to confirm that it will operate if a fault develops on the socket.

9

Alarm Systems and Fault Location

Check that you can remember the following facts from the previous chapter.

Note down the safety precautions that should be carried out before implementing the insulation resistance test. Check back in the book to ensure that you have remembered them all.

On completion of this chapter you should be able to:

- identify manual methods of operating an alarm system
- identify automatic methods of activating an alarm system in an emergency
- identify, from block diagrams, a security system and a space heating control system
- describe, with the aid of block diagrams the operation of a security system and a space heating control system
- identify the types of sensors used with a fire detection system (automatic and manual)
- identify the need for a logical approach to fault location
- describe how to fault find in the following situations
  - earth continuity
  - ring final circuits
  - insulation resistance
  - polarity
- identify the factors involved in rectifying faults

Part 1

Alarm systems

We will now examine principles and control components that form the basis of any alarm system. Even the most complex of control systems, such as those used in building energy conservation, are based upon similar principles. To begin with we will look at the methods we can use to operate a simple alarm system. The function of any alarm system is to give a warning in the event of a change in conditions, and typical examples are fire or intruder alarms. The alarm can be triggered either by some manual means such as the operation of a switch, the principle used in a fire alarm break glass unit, or by an automatic device such as a smoke detector.

In its most simple form an alarm may consist of a switch or push button such as may be located below the counter of a bank. The switch is operated manually by the teller in the event of an attempt being made to rob the bank. Similarly a fire alarm system may consist of manually operated break glass points located adjacent to the building exits.

It is generally accepted that the best form of detection system for fire or intruders is the human, with senses of hearing, sight, smell and touch. However, it is not practical to have all buildings manned 24 hours a day to protect them against the risk of fire or intruders. Protection is usually achieved by the use of automatic devices which operate when there are no humans present to raise the alarm. The type of device used will depend on the particular conditions that prevail and the protection requirements.

Fire alarm systems

We shall first consider systems for fire detection and alarm such as that shown in Figure 9.1.

(S)	Smoke detector
(H)	Heat detector
▢	Break glass
⏚	Alarm bell

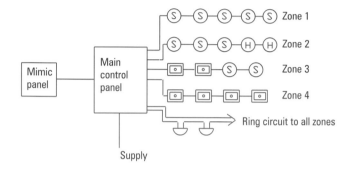

Figure 9.1

This system consists of
- main control panel
- mimic panel
- detectors of different types
- alarm sounders

Let's consider each item in turn.

Main control panel

The main control panel which forms the centre and "brains"of the system. Modern alarm panels are in fact computer processor units and will include components to ensure a battery back up power supply, constant monitoring of all circuits and a fault indication system. In the event of any of the sensing devices operating the main control panel will cause the sounders to operate, indicate in which area the device has operated and can perform any number of functions via telephone lines and direct links. The design and production of these control panels are specialist jobs and they are generally purchased preprogrammed to perform the functions specified by the client.

A system for a large building may include a mimic panel which simply mirrors the information shown on the main panel. The mimic is used to relay the information to a particular location, such as a security office or caretaker's accommodation so the system can be easily checked.

Detectors

We are more concerned with the devices connected to the main panel which provide the signals to operate the system. You will see that these devices are divided up into zones, and this is a common practice to enable accurate location of devices. On large installations it is not practical to simply wire all the detectors on a single circuit. Both the tenant and the fire brigade need to know the location of a fire or fault to within a relatively small area. Zoning has the added advantage of allowing the system to continue to function on all other zones in the event of one developing a fault or cable damage.

There are three basic types of detector shown and these may be mixed within a zone providing they are of compatible types.
- Manual break glass
- Smoke detectors
- Heat detectors

Manual break glass

The break glass unit is a manually operated switch, a typical example of which is shown in Figure 9.2. It is intended for operation by the occupants whilst the building is in use. The human operator can sense smoke, heat, flame or sound of fire and sound the alarm by use of the break glass whilst evacuating the building.

Figure 9.2   A manually operated break glass unit

Smoke detectors

The smoke detector is generally one of two types, either ionised chamber or optic chamber. Both operate by sensing the presence of smoke particles within the detector head. They must therefore be placed in locations where smoke will begin to collect initially to give the earliest possible warning.

Ionised chamber type

This comprises two ionisation chambers with a low energy ionising radiation source. One chamber is sealed, the other is open so that air can flow through it. Under normal conditions ionised molecules of gases flow freely across the ionisation chambers causing a small current to flow. Ionised smoke particles impede the flow across one chamber so reducing the current in that chamber. When the current has dropped to a predetermined level, the alarm circuit is triggered.

Figure 9.3   *Smoke detector*

Optic chamber type

A concentrated pulse of light is projected across a matt black chamber. A photosensor is fitted so that under normal conditions little or no light shines onto it. When smoke enters the chamber, light from the source is scattered by reflection and some falls on the photosensor. This produces a small signal which is amplified and used to activate the alarm.

Heat detector

The heat detector may be a bimetallic type but most are now thyristor operated. As the temperature rises the resistance of these devices changes. At a preset temperature sufficient change of resistance is produced to cause the device to operate an electronic switch causing the alarm to sound.

Figure 9.4   *Heat detector*

The selection of the type of device will be dependent on the building's use. A smoke detector will not sense a fire producing little or no smoke but high temperature. Similarly a heat detector will not sense a fire producing a lot of smoke but low heat.

Alarm sounders

On simple systems it is common to wire the sounders on a single circuit throughout the building to alert everyone to the presence of a fire and evacuate the building. The sounders are generally wired as a ring circuit to ensure reliability and security of the sounder system. Modern alarm systems, with micro processor monitoring and sensing will allow all types of devices, both detectors and sounders to be wired on a common

circuit and allow the zones to be programmed. When a fire is detected, the alarm can be sounded at different levels in each zone. On larger buildings this will allow the evacuation procedure to be programmed and controlled to enable all personnel to exit the building safely and minimize the risk of panic. This is generally done by sounding the alarm in the immediate area of the fire continuously and immediately, the adjacent areas would have the alarm sounded, with a different tone and usually not continuously, to make the occupants aware that they will need to prepare to evacuate the building. After a predetermined period the adjacent areas have the continuous alarm and the areas adjacent to them have the warning and so on until either the fire is controlled and the alarm silenced or the building is completely evacuated.

Figure 9.5   *Electronic sounder*

The use of the micro processor and sophisticated control and sensing equipment has resulted in the production of "intelligent systems" which carry out self diagnosis and advise of problems, monitor the building and equipment, control the evacuation of the building, shut down air circulation systems inform the fire services and so on. This is a far cry from the simple break glass switch against the exit door but based on the same principles.

Intruder detection systems

This basic principle of operation also applies to a security alarm system. In this particular case we are generally concerned with the protection of property or contents whilst they are left unattended. Because of this "unmanned" factor, the majority of sensors used are for automatic operation. Figure 9.6 shows a block diagram for a simple security alarm system.

▦	Door contact
▤	Window contact
⊙	Panic button
⊩	Infra-red heat sensor
◁)	Movement sensor
▷	Sounder

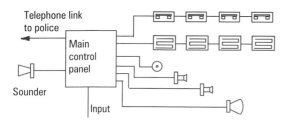

Figure 9.6

Again we can consider each component in turn.

The main control panel
This is similar to the control panel for our fire alarm system, generally a computerised control centre. It will monitor the condition of each circuit, notify any faults that occur and in the event of an intruder it will operate the necessary outputs to sound the alarm

Panic button
Only the panic button is manually operated and its function is similar to an ordinary bell push, once operated the alarm goes into operation.

Door contact

The door contact, shown in Figure 9.7, often consists of a simple magnetic switch which is operated when the door is opened.

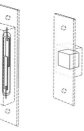

Figure 9.7   *Magnetic door switch*

Window contact

The window switch may be a magnetic switch similar to that fitted to the door or it may be a thin wire attached to the inside of the glass, shown in Figure 9.8. Any attempt to break or cut the glass to obtain access will cause the wire to break and trigger the alarm.

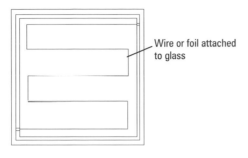

Wire or foil attached to glass

Figure 9.8   Window contact

Infra-red heat sensors

Infra-red heat sensors are often used within a building and they detect the amount of heat radiated by objects within their range. Any person entering the room will trigger the devices due to the change in temperature caused by the heat produced by the human body.

Movement sensor

The movement sensor is a similar device only this operates on a reflected infra-red light signal. A person moving within the range of the device refracts the light source, which is not visible to the naked eye, causing the device to operate and trigger the alarm. A similar device is used to control auto-on security/outside lights and these are generally known as P.I.R. (passive infrared) detectors

All these devices are intended to protect against intrusion and so require adjustment to prevent nuisance tripping by small animals such as cats, dogs and birds. Once operated all these alarm systems will require some form of key reset. This may be by use of a key or a number code or smart card. Operation of any device will cause the alarm to sound, simply resetting the device will however not cancel the alarm signal.

These are common types of alarm systems. In addition there are alarm systems to guard against ingress or loss of water, air pressure, temperature - in fact the list is almost endless. Many of these have specialist applications but all operate on the same basic principle of a sensor supplying information to a processing unit which in turn operates the alarm sounders and the like.

Try this

Using the layout of the building for the project, list the location and type of device you would install to protect the building against intrusion.

Space heating systems

In addition to alarm systems there are numerous control circuits used for a wide range of applications.

Let us look at a simple space heating control system as a typical example and as before we will consider it as a block diagram. Figure 9.9 shows such a diagram for a typical space heating system, including the provision of hot water from the same boiler.

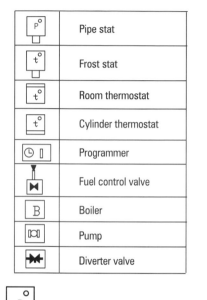

	Pipe stat
	Frost stat
	Room thermostat
	Cylinder thermostat
	Programmer
	Fuel control valve
	Boiler
	Pump
	Diverter valve

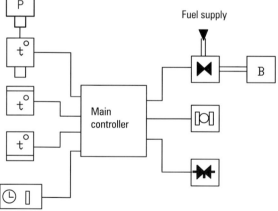

Figure 9.9   Block diagram for a space heating system

Contained within the space heating system we have a number of sensing components, a control unit and some output devices.

In the first instant let us consider the sensing devices.

Thermostats

In the main these will be measuring temperature and our particular system has three such devices or thermostats. The room thermostat is required to monitor the air temperature within the space being heated. This will signal for more heat if the air temperature falls below the set level and signal shut off when the required level is achieved. For domestic applications these are generally bimetal switches as shown in Figure 9.10.

Figure 9.10   Room thermostat

There is also a thermostat monitoring the temperature of the domestic hot water. This can be either a bimetal device similar to the room thermostat or an invar rod in a brass tube immersed in water, as used with an electric immersion heater. Its function is the same as that of the room thermostat, only monitoring water temperature.

A frost stat is generally mounted outside a building and acts as an override device calling for heat should the outside temperature fall below the set level. It will generally be used to override any time control to enable the system to maintain a suitable temperature should there be a sudden drop in the outside temperature. It is particularly useful for office buildings where a day omitting time switch may be used so that the heating will not run over the weekends, the frost stat will prevent freezing up of pipes and such like during this period.

A frost stat is often used in conjunction with a pipe stat installed for energy conservation. The pipe stat controls boiler operation by measuring the return water temperature to the boiler.

Programmer
The other input device shown on our block diagram is the programmer. This is in effect a glorified time switch although extra refinements may be included to allow for various combinations of space/water heating, day omission and the like.

Output devices
Next we shall consider the output devices connected to our system. There are two main output devices the pump and the boiler itself. The function of the pump is to circulate water around the system through the boiler, radiators and hot water cylinder coil. The function of the boiler is to provide heat to raise the temperature of the water being circulated.

Control devices
This leaves us with the control devices. There is the main control unit which receives all the data from the input devices and operates all the other control and output devices. As you can see, it is very similar to the main control panel for our

alarm systems in its function. The diverter valve is operated by the controller to ensure the pumped water is delivered to the priority requirement. This can be set to give priority to space heating over water heating or vice versa. Once one of these reaches its desired level the valve can be driven to divert all the pumped water to the remaining service - thus maximising the boiler output.

The fuel control valve, as its name suggests, controls the flow of fuel to the boiler. If we assume for this example that a gas boiler is being used then when both air and water temperatures have reached the required levels the boiler will be shut off. In this event the supply of gas to the burner must also be switched off.

The controller carries out this function via the fuel control valve. The valve is usually fitted with a thermal link which will melt and release a spring loaded lever shutting off the fuel to the boiler should the temperature within the boiler become too high, irrespective of any other parameters or controls.

Points to remember ◀ – – – – – – – – – –

Switches and push buttons may be used to provide manual operation of alarm systems. Automatic control devices are the most common and cover a wide range. Most control and alarm systems comprise a number of sensors or input devices, a central control unit and a variety of output devices covering a wide range of application.

Try this
Using the layout of the building for the project, produce a block diagram showing the heating controls and state where you would locate each item

---

Short answer questions

1. Draw a block diagram of a gas fired warm air heating system for a domestic dwelling, including water heater and frost protection.

2. List the automatic detectors that may be used for a fire detection system and state one advantage and one disadvantage for each.

Part 2

Fault location

We have considered the process of inspection and testing and the need to record the results of the test and in the process we are going to come across circuits, cables or equipment which have faults. In some cases the customer may report a fault with a circuit or piece of equipment which they want repaired. It is important to consider the difference between the faults we identify on the electrical installation and those referred by the customer. In general the customer reports a problem when a circuit or piece of equipment fails to operate correctly or when it ceases to operate at all.

The effect of a circuit without a suitable Z_s may only become apparent to the user when it is too late and someone has suffered an electric shock. The purpose of the initial and periodic checks on installations and equipment is to ensure that no such "hidden dangers" exist. Any that we do find should be repaired, in the case of the initial inspection and test or, in the case of a periodic inspection, brought to the attention of the person responsible for the installation. Of course the Electricity at Work Regulations place a responsibility on us to ensure that we do not leave an installation in an unsafe condition, so even on a periodic inspection we may need to take some emergency remedial action if a dangerous situation is found.

The process of fault location requires some logical thought and a process which allows us to investigate the fault in the simplest way. Fetching a double extension ladder, climbing to a high outside light, replacing the lamp only to find that someone has switched the circuit off is a typical example of beginning the remedy without a logical approach and checking the simple things first. Dismantling a domestic oven to test the elements as it is not working, only to find the customer has switched the oven to timer is another typical example. When faced with a fault we need to approach it logically and methodically, starting with the obvious and progressing through a stage at a time.

On some occasions it may be enough to identify and report that there is a fault but most customers will require more information and ask for the fault to be located and eventually rectified. When rectifying faults consideration should be given to such factors as cost, resources, safety (of both personnel and the installation) and minimising the time the installation or circuit is switched off to avoid disruption.

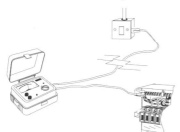

Figure 9.11

In this lesson we shall be looking at the location and rectification of faults on the fixed wiring of the electrical installation. In particular we shall consider the location of faults found whilst carrying out the tests for

- Earth continuity
- Ring circuit continuity
- Insulation resistance
- Polarity

First we need to consider a logical approach to our investigation process so that each stage can be examined and we progress one step at a time.

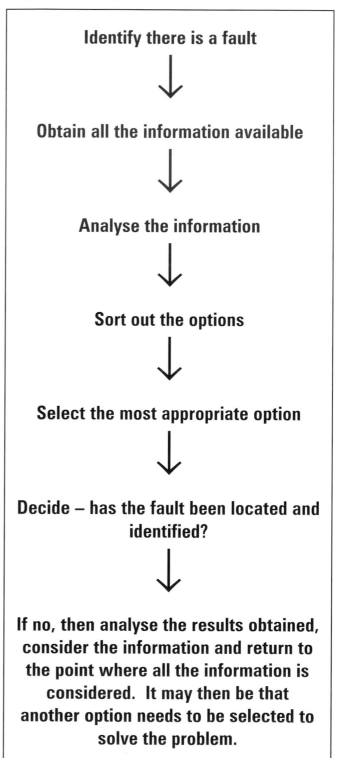

118

Now let's look at each step in more detail before we apply them to the fault finding process.

Identify if there is a fault

It is important that we establish what the symptoms of the fault are in order that we can determine whether a fault really exists. If the person providing the information is not familiar with the normal operation of a particular item of equipment then it is possible that the perceived fault is the normal operation. For example "My fan heater keeps cutting out and if I leave it alone it eventually comes on again for a little while and then cuts off again". It may well be that the fan heater is controlled by a thermostat and the "fault" is the normal operation of the thermostat.

So first make sure there is a fault to be investigated.

Obtain all the information available

The information available will include
- any information on events leading up to the fault
- original drawings and specifications, where available
- manufacturers' details, Regulations and other British Standards
- knowledge and understanding of the circuits and protection and control devices involved
- personal and others' experience and expertise
- instrument readings obtained so far
- visual inspections

Analyse the information

Collate the information and list any appropriate actions that may be considered relevant. Make sure that all information is complete, accurate and recorded so that it can be made available as required. Put to one side, but do not discard, any information which appears irrelevant or misleading.

Sort out the options

There may be more than one possible course of action which could be taken and some may be easier and quicker to try than others. If an option requires help that is not available then that may need to be put to one side until help is obtained.

Select the most appropriate

Once a course of action has been chosen, it should be carried out and the results analysed. If some disruption is expected then the people affected should be informed and permission obtained before proceeding.

Has the fault been identified?

If it has been identified, can it be put right? It is important that when one fault is located and corrected, another is not introduced. In order to ensure that this does not occur the original tests will usually have to be repeated before the circuit can be put back into use.

Rectification

Each option available for correcting a fault should be assessed taking into account:

cost – repair or replace components?
is a stand-by system required?

resource constraints –
if a replacement is required is it readily available?

safety of personnel –
do other trades on site, colleagues or clients need to be consulted with regard to isolation?

safety of installation

area of responsibility –
is there someone with sufficient authority available to authorise the repair?

For our first example let's consider a circuit which is being tested for earth continuity.

Earth continuity

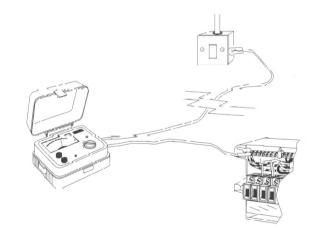

Figure 9.12

A lighting circuit wired in PVC insulated and sheathed cable is tested from the distribution board to a metal switch box with a long lead and shows no earth continuity.

This shows a definite open circuit, but before going into the fault location routine the test instrument and its leads should be checked.

Assuming these to be in working order we must now attempt to trace the fault.

Obtain all the information available

As this is one of a series of earth continuity tests the other results should be studied. A switch point is often the end of a cable run - we need to know if it is in this case or if it goes onto another switch, such as in a two-way system.

Analyse the information

If the test results for the rest of the circuit are satisfactory then the fault must be related to this one cable run.

Sort out the options

In this case there are two probable options

- take off the switch and see if a wire has become disconnected
- trace out the switch cable to the ceiling rose and check the connections

Select the most appropriate option

The most appropriate may be the easiest option so in this case the removal of the switch is worth a try.

Has the fault been identified?

If the fault is not at the switch then the other option remains open. In either case it should be possible to identify and repair the fault without alteration to the circuit.

Try this

Earth continuity

When checking the continuity of a steel trunking cpc, 30 m long, a value of 100 Ω is obtained when the test is carried out from the earthing terminal in the main distribution board to the metal case of a motor isolator/starter supplied through the trunking system.

Explain the procedure you would adopt to identify the fault and state what the likely cause may be.

Ring final circuits

On completion of an installation using single insulated PVC cables installed in PVC conduit and trunking we are carrying out the initial inspection and testing of the installation before it is commissioned. When carrying out a ring circuit continuity test, using the method described earlier in this book, the following results were obtained.

At the distribution board with the ends of the ring circuit, which is wired using 2.5 mm² conductors throughout, disconnected :

Phase to Phase	*0.6 Ω*
Neutral to Neutral	*0.6 Ω*
CPC to CPC	*0.6 Ω*

At the distribution board with the alternate ends of the ring connected together: 0.3 Ω

As we begin to carry out the confirmation that the circuit is actually a ring the readings obtained at each socket begin quite low and gradually increase as we get further away from the distribution board. As we get further round the ring circuit the readings reach a maximum and then gradually decrease as we get closer to the distribution board again.

Obtain all the information available

The information for this circuit could include an as fitted drawing of the installation detailing the cable runs. A visual inspection of the installation could be carried out, although correct installation of conductors is not possible with a conduit and trunking installation by visual inspection. What we need to consider in this case is our own understanding of the test and our knowledge of the effect of cable resistance.

The information we have in relation to this fault is

- the circuit appears to be wired as a ring circuit as the initial continuity test shows end to end readings of similar values for each conductor
- the readings obtained at each outlet follow a trend, the further they are from the distribution board the higher the reading becomes
- the resistance of a conductor increases in direct proportion to its length

Analyse the information

Figure 9.13 shows the configuration of the ring circuit and the requirements for the tests. If we consider the circuit as a series of resistors then it would appear as shown in Figure 9.14 and if connected in this way the ring circuit test would produce a similar reading at each point on the circuit. So we know that this is how the circuit should be arranged and that our circuit appears to be arranged differently, and we know that the effect is that the circuit resistance increases towards the centre of the circuit and reduces towards the distribution board.

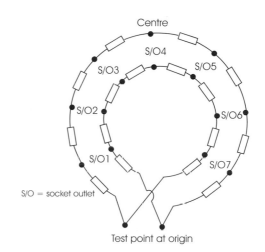

Figure 9.13

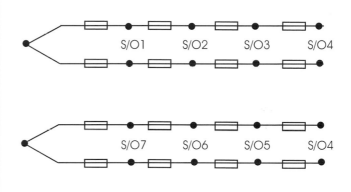

Figure 9.14

If we consider the effects of the resistance of the cable, we know that this increases with length. In that case our circuit appears to get longer to the centre and then shorter as we move towards the ends. This indicates that the circuit has a higher resistance in the centre and if we consider the effect it would appear that the circuit comprises two branches which are approximately equal and show similar characteristics. Figure 9.15 shows the circuit diagram equivalent of this, but we know that the circuit was tested and ring circuit continuity, phase to phase and neutral to neutral, appeared to be correct. It appears that the final connection of the circuit for this part of the test is incorrect.

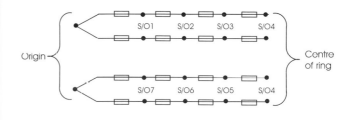

Figure 9.15

So what could produce this effect and how could this have occurred?

With a circuit wired in PVC single core cables in conduit and trunking it is not always easy to identify the ends of the ring circuit to ensure that the opposite phase and neutral conductors are connected together for this test. If we have connected the phase and neutral conductors of each end of the ring circuit together then what effect will this have on our circuit? The circuit we are testing will be a circuit exactly like that shown in Figure 9.15.

Sort out the options

In this particular case we have, by considering and analysing the information with the benefit of our understanding of the circuit characteristics, come to a single likely option. The options available to us are

- remove the sockets and trace out the circuit
- trace the ring circuit from one end working to the other
- take of a socket near to the centre of the ring and separate the circuit into sections
- connect the phase and neutral conductors together in the opposite configuration

Select the most appropriate

From the information we have and the analysis we have carried out the most probable scenario is that the circuit is not connected with the opposite ends of the conductors placed together. So the remedy for that is the simplest, quickest and easiest to carry out so that is the one to try first.

On changing the connections and repeating the test we find that the test at each socket results in the same value being obtained at each socket outlet.

Has the fault been identified?

In this case the connections made at the distribution board were incorrect and the opposite ends of the circuit had not been connected together. Changing the configuration of the connections rectified the problem. The process of considering what we know about the circuit, the properties of conductors and the expected results of the test enabled us to readily identify the problem and rectify the fault without unnecessary dismantling of the circuit.

On this occasion the fault was as a result of incorrect connections at the time of carrying out the test. Remember that the process of identifying a problem with a circuit will also include the checking of instruments and the connections used during the test process as well as the circuit under test.

Remember

If the values of resistance, obtained at each socket during the confirmation of ring continuity, rise as the sockets get further away from the distribution board, and are lower as they get closer to the distribution board, it is most probable that the opposite ends of the ring circuit have not been cross connected.

Try this

Ring final circuit

A ring circuit continuity test was started at the distribution board but when the following results were obtained the test was stopped.

Phase to phase   0.7 Ω
Neutral to neutral   open circuit

It was confirmed that the neutral conductors were those associated with the circuit.
Explain the procedure you adopt to identify the fault and state what it may be.

Insulation resistance

An insulation resistance test on the supply cables to earth of a single-phase 10-way distribution lighting board gives a reading of 0.3 MΩ.

As the minimum acceptable resistance is 0.5 MΩ then the reason for the reading of 0.3 MΩ must be found.

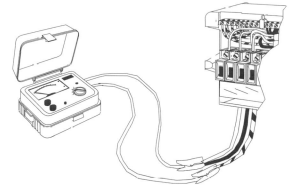

Figure 9.16

Obtain all the information available

The information available from the chart at the distribution board tells us that there are 10 circuits in use using a mixture of tungsten and fluorescent lighting. Two of the circuits are supplying stores approximately 20 metres away from the distribution board. The other circuits supply offices, toilets, washrooms, work areas and a reception.

We must also consider what we know about the nature of the test, the expected result and the factors which affect the test results. In the case of the insulation resistance test we need to consider items of equipment which may have a leakage to earth and ensure that these are disconnected from the installation. We need to remember that items of equipment which are supplied through single pole isolators and protective devices will still have the neutral side connected to the main neutral terminal.

We are also aware that the testing of insulation resistance is effectively testing a number of resistances connected in parallel. We also know the effect of resistances in parallel is to produce an overall resistance which is lower than any of smallest of the resistors connected in the parallel group. When a number of circuits are connected in parallel to the distribution board then the overall resistance measured at the tails supplying the board will be less than the individual resistance of any one circuit.

Analyse the information

This combination of circuits could lead to a number of reasons for the low test result, including

- a faulty component in a fluorescent luminaire
- damp within a luminaire or switch in the washroom area
- number of outgoing circuits
- damage to the insulation of a cable

Sort out the options

When we consider the options available to us at this stage, the first choice must be an option to determine whether there is a fault on one or more circuits. In order to establish this we have a number of alternatives available We can

- disconnect each circuit in total from the distribution board and test each one in turn.
- switch off each MCB and test the board again and establish if the fault is removed
- disconnect each neutral conductor in turn to establish which circuit the fault is on
- establish by testing the phase and neutral conductors separately to earth whether the fault is on both or just one of the live conductors
- establish whether there are any items of fixed equipment connected to the installation

Select the most appropriate option

If we consider the options we have at this time it would appear that we need to find out a little more before we can make any meaningful investigation. The most straightforward action which will give the most useful information is to find out if there is a fault on both phase and neutral conductors.

We have carried out the insulation resistance test between the phase and neutral conductors connected together and earth and have established that the result is below the minimum acceptable value. There is one simple test we can do to find out whether this fault exists on both live conductors or just in one, so the first action would be to carry out the test again between the phase and earth and then between neutral and earth and record the results.

Has the fault been identified?

In this particular instance we establish that the fault is on the phase conductor, but we have not established which circuit the fault is on and so we need to go back to the options again with this further information and see whether there are any further options to consider having carried out some further tests.

Select the most appropriate option 2

We have established that the fault is on the phase conductor only and the previous list of options will still apply. The most appropriate action would be to determine which circuit or circuits the fault is on. The most effective method of doing this is to switch off all the circuit-breakers and then switch each one on in turn and carry out the insulation resistance test to establish which circuit or circuits the fault is on.

Tip : Do not stop carrying out the test once the first fault is located as there may be more than one faulty circuit. Switch off each circuit after completing the test in order to identify each circuit on which there may be a fault.

Has the fault been identified?

Once the faulty circuit or circuits have been identified we need to carry out further tests to locate the actual fault and determine the remedial action required.

Insulation resistance

An insulation resistance test on a ring final circuit gives a resistance of $10\,\Omega$.

Explain the procedure you adopt to identify the fault and state what it may be.

Polarity

When carrying out a polarity test on a number of socket outlets in a ring final circuit, two were found to have reversed polarity.

As only two outlets have reversed polarity, the fault appears to be at the individual sockets.

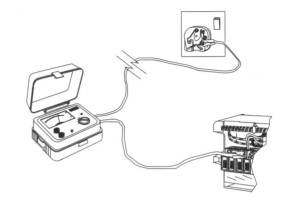

Figure 9.17

Obtain all information available

A check should be made to see if the two socket outlets are run consecutively on the ring.

Analyse the information

If the two sockets are connected one at each end of a single section of cable, it is just possible that it is connected incorrectly. It is far more likely that the two sockets have been incorrectly connected.

Sort out the options

Remove each of the sockets with reverse polarity in turn and reconnect as necessary.

Has the fault been identified?

By removing both sockets the fault can be identified and rectified at the same time.

Try this

Polarity

When carrying out a polarity test on a lighting circuit, prior to the connection of the supply, the switches are found to be in the neutral conductor. A visual inspection shows that the cables are currently connected at the distribution board. The circuit is wired in PVC insulated and sheathed cable in a 3-plate system.

Explain the procedure you would adopt to identify the fault and state what it may be.

Points to remember ◀ — — — — — — —

Fault location can be a very time consuming task. To keep the time and frustration down it is important to develop a logical approach. If used correctly, this approach can help to build up confidence when dealing with any problem.

Plan a logical approach to diagnose and rectify causes of electrical fault.

> Identify there is a fault
>
> Obtain all the information available
>
> Analyse the information
>
> Sort out the options
>
> Select the most appropriate
>
> Has the fault been identified?
>
> Select, and action, the preferred rectification option.

Symptoms of the fault may include items such as:

> Total shut down
>
> Loss of power
>
> Short circuit
>
> Low impedance
>
> Power surge
>
> Plant failure
>
> Component failure
>
> Damage caused by negligence or malicious damage

Some situations require special precautions to be taken.

For example:

> Electronic components may be damaged by insulation resistance test high voltages.
>
> Some components use electrostatic sensitive devices that can be damaged by static electricity and must therefore be handled in the correct manner.
>
> The ends of fibre optic cabling can transmit unexpected light signals which may cause damage to sight. Fibre optic cabling can also be damaged by overbending.

Try this

A piece of equipment has been found to be faulty. What are the factors that would influence the decision as to whether the equipment should be repaired or replaced.

End test

Multi-choice questions

Circle the correct answers in the grid below.

1. A ring final circuit is wired in 2.5 mm^2 cable and has a total length of 30 metres. The resistance, when measuring the continuity of the phase conductor, should be approximately
    a. 30 Ω
    b. 3 Ω
    c. 0.3 Ω
    d. 0.03 Ω
2. Where steel conduit is used as the circuit protective conductor a continuity test should be carried out using
    a. > 500 V a.c.
    b. less than 50 V a.c.
    c. 500 V d.c.
    d. 25 V d.c.
3. The resistance between two adjacent central heating pipes is found to be 1 MΩ. This means that
    a. the resistance is satisfactory
    b. the two pipes need bonding together
    c. a test should be carried out to earth
    d. the main equipotential bonding is faulty
4. An enclosure that restricts the access of objects around barriers of 1 mm diameter conforms to
    a. IP1X
    b. IP2X
    c. IP3X
    d. IP4X
5. An earth electrode resistance test can be carried out using only
    a. a continuity tester
    b. a milliohmmeter
    c. a high current d.c. tester
    d. an earth ohmmeter
6. If an RCD is installed on a TT system, the acceptable value of earth electrode resistance multiplied by the trip rating of the RCD must not exceed.
    a. 50
    b. 100
    c. 150
    d. 500
7. A test that conforms to IP2X requires that a probe of specified diameter cannot penetrate an enclosure. The diameter of the probe is
    a. 1 mm
    b. 2 mm
    c. 3 mm
    d. 4 mm
8. On an insulation resistance test on an installation which of the following readings is acceptable?
    a. 0.25 MΩ
    b. 0.35 MΩ
    c. 0.45 MΩ
    d. 1.05 MΩ

9. A polarity test is carried out to confirm
    a. all connections are electrically and mechanically sound
    b. all switches and control devices are in the phase conductors
    c. that circuits will function as intended
    d. that all socket outlets are connected to a ring final circuit
10. An RCD with trip current of 30 mA is used to protect part of a TT system. The maximum resistance of the earth electrode for this installation must not exceed
    a. 1700 Ω
    b. 150 Ω
    c. 50 Ω
    d. 1.7 Ω
11. To test that an RCD will operate under fault conditions satisfactorily, tests are carried out at 50% and 100% of the rated tripping current and 150 mA. The trip should operate on
    a. only 50%
    b. 50% and 100% only
    c. 100% and 150 mA only
    d. all three tests
12. The impedance on an earth fault loop is measured against a maximum test current of
    a. 75 A
    b. 35 A
    c. 25 A
    d. 12 A
13. When testing portable electrical equipment the earth bond test must be carried out on all equipment
    a. Complying with Class 1 specification only
    b. Complying with Class 2 specification only
    c. that does not have a conductive enclosure
    d. designed for use with a 2 core flexible cable
14. A 24 V a.c. r.m.s. supply when displayed on an oscilloscope will have a peak to peak voltage of
    a. 24 V
    b. 48 V
    c. 68 V
    d. 79 V

Answer grid							
1 a	b	c	d	8 a	b	c	d
2 a	b	c	d	9 a	b	c	d
3 a	b	c	d	10 a	b	c	d
4 a	b	c	d	11 a	b	c	d
5 a	b	c	d	12 a	b	c	d
6 a	b	c	d	13 a	b	c	d
7 a	b	c	d	14 a	b	c	d

Short answer questions

1. A visual inspection has to be carried out on a domestic installation wired in PVC insulated and sheathed cable. Make up a list of items that will need to be inspected.

2. An insulation resistance test is to be carried out on a lighting distribution board. Explain the procedure that must be adopted before any tests can be carried out.

3. The test button is pressed on an RCD and the device does not switch off. Explain what needs to be checked to determine if the RCD is faulty.

Answers

These answers are given for guidance and are not necessarily the only possible solutions.

Chapter 1
p.4 Try this: 718.8 V, 400 V, 190.5 V

p.9 Try this: 86.6 A, 415.7 A, 1732 A or 1.732 kA

p.10 Try this: 1. Single pole; 2. Triple pole; 3. Triple pole with bolted neutral; 4. Double pole

p.13 Try this: 90%

p.16 Try this: A

p.17 MC 1. b; 2. c; 3. b; 4. d; 5. c; 6. b; 7. a; 8. b; 9. b; 10. c

p.18 SAQ 1.i) An overload current occurs in a circuit that is electrically sound. 1.ii) A short circuit current is a fault current within the circuit or components. This overcurrent occurs when two live conductors, with a difference in potential under normal conditions, connect.

SAQ 2. Relevant charts and drawings may be useful to identify where and what type of circuits, equipment, means of isolation, characteristics of components/devices are incorporated in the system.

SAQ 3. 109.77 A. Options which may be considered by the supplier are diversity and maximum likely demand.

Chapter 2
p.22 Try this: Type B: domestic installations, industrial small power circuits. Type D: motor circuits, large inductive load circuits (such as high bays)

p.23 Try this: 0.05 s

p.28 Try this: 1. T – supply connected directly to earth, N – exposed metalwork connected directly to earthing point of supply, C – some part of the system has combined neutral and earth, S – some part of system has neutral and earth separate; 2. earth fault loop; 3. phase and earth; 4. low; 5. earth electrode

p.32 MC 1. b; 2. d; 3. b; 4. a; 5. d; 6. d; 7. b; 8. b; 9. c; 10. a

p.34 SAQ 1. **Semi-enclosed fuse** – advantages: relatively cheap, easily repaired, fairly reliable, easy to store spare wire, easy to see when fuse blown; disadvantages: easily abused, high fusing factor, precise conditions for operation, do not cope well with high short circuit currents, wire can deteriorate.

HBC – advantages: have a low fusing factor, ability to break high currents, reliable, accurate; disadvantages: expensive, spares costly, spares take up space, care needs to be taken when replacing to ensure correct one.

MCB – advantages: only need resetting, setting cannot be adjusted, can cater for short circuit currents, discrimination between harmless transient overloads and short circuit currents, easy to identify breaker has tripped; disadvantages: expensive, mechanical, cannot be used if the short circuit current is in excess of their short circuit rating.

SAQ 2. 0.01 seconds or less.

SAQ 3. Test button completes the circuit containing the resistor which causes an imbalance in the flux in the coil of the RCD. The sensing winding then trips the double pole switch and cuts off the supply.

SAQ 4. See Figure 2.11

Chapter 3
p.39 Try this: 12.077 A

p.39 Try this: 45.1 A

p.41 SAQ 1. Calculate the power required if every piece of equipment installed was switched on, for example floor warming installation. Calculate the power required when diversity is applied, for example a lighting installation in a small hotel.

SAQ 2. I_b – the design current drawn from the supply under normal conditions.

SAQ 3. The number and types of conductor, the type of earthing arrangement available, the nominal voltage, the nature and frequency of the current, the prospective short circuit current at the origin of the installation, the type and rating of the overcurrent protective device, the suitability of the supply, the earth fault loop impedance.

p.42 SAQ 4. 16.75 A

SAQ 5. 0.71

Chapter 4
p.49 Try this: 21.3 A

p.52 SAQ 1.(a) increases resistance in cable – less current can flow (b) cable size needs increasing to reduce heat generated in cable

SAQ 2. All cables will produce heat   less heat dissipation – less current can flow

SAQ 3. 13.2 A

Chapter 5
p.56 Try this: 4 mm^2

p.58 SAQ 1. ambient temperature, grouping, thermal insulation, semi-enclosed fuse used

SAQ 2. max 8 mV/A/m

SAQ 3. 4 mm^2 at 10mV/A/m

SAQ 4. max load 20.689 A

Chapter 6
p.62 Try this: 1. 24.20; 2. 14.82; 3. 10.49

p.68 SAQ 1. 0.4137 mm^2

SAQ 2. 454.58 A

SAQ 3. 36.99 mm^2

SAQ 4. (a) 980 A; 4(b) 0.6 s

Chapter 7

p.71 Try this: Green sleeving omitted on 2 no. cpc, green sleeving too short on 1 no. cpc, exposed conductor outside terminal on 1 no. neutral, exposed conductor outside terminal on flex, cable clamp not used on neutral flex.

p.73 Try this: **Safety Electrical Connection – Do Not Remove**

p.75 Try this:

1. Sports centre toilet area
- suitability of installed equipment for the environment
- suitable equipment installed
- equipment suitably located
- is there any need for any supplementary bonding and is this installed if required
- signs of corrosion or deterioration of either equipment or wiring systems
- signs of damage to equipment and wiring systems
- rating of local protective devices suitable for the purpose (fuses in spur units etc.)
- In addition other items from listing in BS 7671 Part 7

2. Dental surgery reception area:
- suitability of equipment installed for the environment
- suitable equipment installed
- suitable location of installed equipment
- signs of deterioration of either equipment or wiring systems
- signs of damage to equipment and wiring systems
- rating of local protective devices suitable for the purpose (fuses in spur units etc.)
- In addition other items from listing in BS 7671 Part 7

3. Boiler room:
- suitability of installed equipment for the environment particularly with regard to the ambient temperature and the need for heat resistant insulation to flexes and cables
- suitable equipment installed
- equipment suitably located
- is there any need for any main equipotential bonding and is this installed where it is required
- is there any need for any supplementary bonding and is this installed if required
- signs of corrosion or deterioration of either equipment or wiring systems
- signs of damage to equipment and wiring systems
- rating of local protective devices suitable for the purpose (fuses in spur units etc.)
- In addition other items from listing in BS 7671 Part 7

4. Large open plan office:
- suitability of equipment installed for the environment
- suitable equipment installed
- suitable location of installed equipment
- signs of deterioration of either equipment or wiring systems
- signs of damage to equipment and wiring systems
- do the requirements of BS 7671 Section 607 apply and has this been recognised in the design
- rating of local protective devices suitable for the purpose (fuses in spur units etc.)
- connection and identification of conductors

- installation of fire barriers
- correct connection of equipment
- protection measures against direct contact in place
- In addition other items from listing in BS 7671 Part 7

p.75 SAQ The need for periodic inspection, its purpose and procedures are detailed in IEE Guidance Note 3. In addition the principal differences between initial and periodic inspection and testing is that for the periodic:
- there may well be some extent and limitations placed upon a periodic inspection and test, which is not permitted on a new installation. For example, on a new installation all of the electrical installation is inspected, tested and verified, on a periodic inspection it is not possible, without causing considerable damage, to inspect those parts of the installation concealed within the building fabric.
- some degree of sampling may be involved during a periodic inspection which again is not permitted with a new installation.
- the extent and limitation placed upon the periodic inspection and test of an installation should be agreed with the client and any interested third party before work begins.
- the sequence of tests carried out is different to that for a new installation and this reflects the fact that the installation is energised and in use.

Chapter 8

p.84 SAQ 1. Continuity of ring final circuit conductors
- ensure that the circuit is securely isolated, all the socket outlets have been installed, no equipment to appliances is connected
- disconnect the cables to the ring circuit at the distribution board or at a socket outlet close to the distribution board
- using a low impedance ohmmeter or continuity tester connect the instrument between the two phase conductors and test to obtain a reading
- record the value obtained (r_1 = phase to phase resistance)
- repeat the test with the two neutral conductors and record the values obtained which should be substantially the same (r_n = neutral to neutral resistance)
- repeat the test with the cpc and record the readings obtained (r_2 = cpc to cpc resistance)
- if the values are not within an acceptable tolerance then the circuit requires some additional investigation to establish the reason
- cross connect the phase and neutral conductors and measure the resistance of the conductors
- using a plug top test at each socket outlet between the phase and neutral conductors
- repeat the process cross connecting the phase and cpc for the ring circuit
- carry out the test at each outlet, between phase and cpc
- record the highest value obtained during this test process as the $R_1 + R_2$ value for the ring circuit

- remove the cross connections and terminate the ring circuit to the distribution board

SAQ 2. Testing of steel cpcs

Equipment:
- ohmmeter capable of reading low values of resistance
- continuity tester
- wander lead
- earth fault loop impedance tester
- high current low impedance ohmmeter

Considerations:
- physical condition of the conduit
- any detrimental influences upon the material, (corrosion)
- risk of sparking at poor joints during the test

p.88 Try this: Any five of: supply isolated – no supply to circuits to be tested, all lamps removed, all equipment normally used disconnected, all electronic equipment disconnected or bypassed, all fuses in place, all switches in ON position, switched circuits tested with switches in each direction unless bridged across during the testing.

p.90 SAQ 1. Before commencing the test procedure it is necessary to ensure that:
- arrangements for an appropriate time are made in order that the lighting may be isolated,
- an alternative light source is available in order that work can be carried out safely,
- suitable access and test equipment is available for the appointed time,
- warning notices are positioned to advise that testing is in progress.

The procedure for isolation, for example ensuring that the lights are isolated from the circuit to be checked, can then be undertaken as detailed on p.12 of this book and in IEE Guidance Note 3.

SAQ 2. Insulation resistance test to newly completed reception lighting carried out in accordance with the procedure detailed on p.86, for an initial insulation resistance test of a circuit.

SAQ 3. Electronic test equipment. The following checks and precautions would need to be undertaken in order to ensure the equipment is not damaged:
- where the switched fused spur is a **double pole switched unit** the equipment may be isolated by switching off the double pole switch
- if the connection to the equipment is via a suitable plug and socket on the unit, often used on computer equipment, then the unit may be unplugged at that point. Remember that any tests undertaken will include the flexes used to connect the appliances
- where the switched fused spur is **not** a double pole switched unit and the equipment is not connected via a suitable plug and socket then it **must** be disconnected from the fused spur units before the insulation resistance tests are undertaken.

SAQ 4. Dented MICC cable. The visual inspection should endeavour to establish whether the outer metal sheath of the cable has been split or whether it is intact. If the sheath is split then the cable will absorb moisture from the air and will ultimately fail and so would need to be repaired or replaced. If the metal sheath is sound then the cable would need to be isolated and disconnected from the supply. The appropriate tests that would need to be carried out would be continuity of all conductors and an insulation resistance test of all cores, both between cores and cores to earth. It would be a good idea to compare the resistance of each core to compare them and so it would not be appropriate to confirm continuity by the use of an audible device but by measurement of the conductor impedance.

p.92 Try this: 1. (a) i) 5.33, ii) 1.85, iii) 1.09; (b) i) 6.86, ii) 2.29, iii) 1.14; 2. (a) i) 7.74, ii) 3.04, iii) 1.92; (b) i) 17.1, ii) 5.22, iii) 1.92

p.98 SAQ 1. Polarity test The tests to confirm polarity are by use of:
- wander lead between the origin and each point on the circuit as described on p.91
- the measurement of $R_1 + R_2$ as described on p.79 of this book and in IEE Guidance Note 3.

Remember that the use of equipment on energised circuits only **confirms** polarity, we need to **test** polarity prior to energising the circuit

SAQ 2. The tests for earth fault loop impedance should be carried out as detailed on p.93 of this book and in IEE Guidance Note 3

SAQ 3. The two methods, use of an earth fault loop impedance test instrument and a proprietary earth electrode test instrument are detailed on pp. 94 and 95 of this book and in IEE Guidance Notes 3. The principle problems which may be encountered include:
- insufficient area and unsuitable conditions to carry out the test using the proprietary electrode test instrument (positioning of test electrodes)
- isolating the outbuilding from the supply authority earthing system, particularly where an armoured cable is used for the distribution circuit
- carrying out "live" tests on an installation without a proven earthing system

SAQ 4. The procedure for carrying out the tests are detailed on p.95 of this book and in IEE Guidance Note 3. The maximum values are 200ms for the $I_\Delta n$ test and 40ms for the $5 \times I_\Delta n$ test as detailed in BS 7671.

p.100 Try this:

	user	formal visual	combined
construction site	weekly	1 month	3 months
commercial kitchen	weekly	none	12 months
school	weekly	none	12 months
hotel	before use	6 months	12 months

p.105 SAQ 1. (a) DC voltmeter, dependent on charger output but generally 0–25-V will be suitable

(b) DC ammeter, dependent on charger but generally 0–5 A will be suitable

(c) An oscilloscope with ranges appropriate to the amplifier being tested, generally in the region of 50 Hz to 20 kHz would be appropriate.

2.

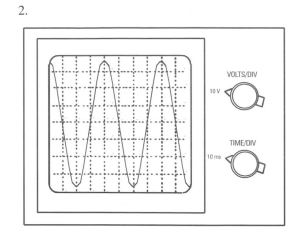

p.105 MC 1. a; 2. b; 3. d; 4. c; 5. b; 6. c; 7. c; 8. a; 9. b; 10. c; 11. d; 12. c; 13. d; 14. d; 15. c; 16. c; 17. d; 18. c

p.117 SAQ 1.

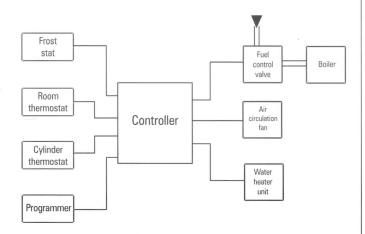

SAQ 2. **Smoke detector:**

– typical advantages:

quickly detects fires which produce smoke,

not affected by changes in temperature

– typical disadvantages:

does not readily detect fires which have high temperatures and produce little smoke,

affected by dust and other airborne particles

Heat detector:

– advantages:

quickly detects fires which produce high temperatures,

not affected by smoke or airborne particles

– disadvantage:

does not detect low temperature fires which produce smoke affected by rise in temperature as result of normal conditions within the location, such as boiler houses and industrial kitchens

Rate of rise detectors:

– advantages:

not adversely affected by areas where a gradual temperature rise occurs normally, such as commercial kitchen locations

not affected by smoke or airborne particles

– disadvantages:

will not detect smoke,

will not detect slow burning fires

This list is not intended to be comprehensive or finite.

p.120 Try this:

The problem is to identify why the measured path is proving to be such a high resistance, this would typically include:

Initially a visual inspection of the trunking route and the joints and terminations should be carried out wherever this is accessible to establish that there are no obvious problems such as

· inappropriate joints

· painted enclosures forming part of the cpc

· isolating packers and spacers, such as mica, used at distribution board/trunking connections

· test made to painted enclosure without suitably low resistance contact.

Once the visual inspection has been completed, or where this is not possible, then the continuity of the cpc path will need to be measured in sections to establish the location of the problem. Generally it would be appropriate to begin at the origin and test sections of the system working towards the point of utilisation. Where access to the trunking is difficult then it may be advisable to check each end of the system for problems before testing the trunking length.

As a guide this may include:

· from the earthing terminal on the main distribution board to the earthing terminal on the motor isolator/starter

· the connection between the main distribution board and the trunking at the point where the two are connected, immediately adjacent to the board

· from the earthing terminal on the motor starter to the point where the enclosure is connected to the trunking system

· from the point on the trunking adjacent to the distribution board to the first trunking joint

· across the first joint

· repeat the process at each joint to ensure the joints are all electrically continuous

If at any point in the process a high resistance is found then the length of cpc tested requires further investigation to establish the cause.

The items identified above in the visual inspection list are the most likely cause of such a fault.

p.122 Try this:

Generally when this occurs the first check should be to ensure that the appropriate neutral conductors are being used for the test. However this was confirmed as correct and the problem remained.

In order to locate the problem the general procedure would be to locate a socket close to the mid point of the circuit and disconnect the conductors. A continuity test on the two "halves" of the ring will generally establish which half of the circuit contains the fault. This process would then be repeated with the faulty half of the ring and narrowing the fault down to a "quarter" of the original circuit. This process would be repeated as often as is necessary to locate the fault whilst minimising the number of sockets that need to be removed.

It is quite likely that a loose terminal resulted in one or both of the conductors becoming dislodged resulting in the open ring reading.

p.124   Try this:

The method used to establish the location of the fault is similar in principal to that used to locate the open circuit neutral problem on page 122. The process of breaking the ring circuit down in that way reduces time, effort and aids speedy location of the problem.

In order to locate the problem the general procedure would be to locate a socket close to the mid point of the circuit and disconnect the conductors. An insulation resistance test on the two "halves" of the ring will generally establish which half of the circuit contains the fault. This process would then be repeated with the faulty half of the ring and narrowing the fault down to a "quarter" of the original circuit. This process would be repeated as often as is necessary to locate the fault whilst minimising the number of sockets that need to be removed.

There are a number of possible causes for this problem the most common being pressure on the insulation of the conductors when accessories are fixed back or at the crutch of terminations where pvc insulated and sheathed cables are used.

p.125   Try this:

As the switches are in the "neutral" conductor and yet the conductors are correctly connected at the distribution board it would be reasonable to assume that one of the ceiling roses has been incorrectly terminated.

Generally the procedure would be to establish whether all the switches are affected and in this instance this would appear to be the case. It would be reasonable, in these circumstances to commence the investigation with a visual check of the first ceiling rose on the circuit from the distribution board. Being a three plate system it is likely that the feed has been connected reverse polarity, at some point, and where all the lights are affected it is likely to be at the first ceiling rose. Where the cable route is unclear a suitable initial procedure would be to carry out a visual inspection of each ceiling rose. It is generally possible to identify the location of the fault by this process.

Where all the switches are not affected it is likely that the fault will be located either at the light where the first reverse polarity is found working away from the distribution board or at the outgoing feed from the previous light.

p.126   Try this:

Some of the typical factors which will need to be considered to determine whether to repair or replace a piece of equipment are

- availability of a) the spare parts & b) a replacement item of equipment
- the period for which the equipment can be out of use
- whether the equipment failure affects other activities or production
- time taken for the repair
- total cost of repair including labour and loss of production against that of a replacement
- the age of the failed equipment, is it near the end of its useful life
- ease of repair
- suitability of the equipment for the task, is an alternative piece of equipment more suitable
- whether the equipment is still required

This is not a definitive or exhaustive list and other factors may need to be considered.

p.127   End Test MC 1. c; 2. b; 3. b; 4. d; 5. d; 6. a; 7. d; 8. d; 9. b; 10. a; 11. c; 12. c; 13. a; 14. c

p.128   SAQ 1. The list of items to be inspected are shown in BS 7671:1992 and IEE Guidance Note 3, those applicable to domestic installations should be included.

SAQ 2. The procedure is explained on p.85 of this book.

SAQ 3. The appropriate checks need to be made to ensure that the conditions required for operation of the device are in place and a test of the RCD carried out. The procedure for carrying out the tests are detailed on p.95 of this book and in IEE Guidance Note 3. The maximum values are 200 ms for the $I_\Delta n$ test and 40 ms for the $5 \times I_\Delta n$ test as detailed in BS 7671.

If the RCD meets the conditions then the device is operating and providing the necessary protection against electric shock. In that case the integral test facility is not functioning and the user is unable to confirm operation periodically as required by BS 7671. The device should be replaced to enable this to be undertaken.

If the RCD does not meet the requirements then it should be replaced immediately as the installation does not have the required protection against electric shock and thus danger could occur to the users of the installation. The installation, or the part which is protected by the RCD, should be isolated until such time as the replacement has been fitted and a suitable level of protection has been provided against electric shock.

Appendix

Contents

Introduction

This project has been designed to be used in conjunction with this book "Stage 1 Design".

The plans show a small factory and office with the electrical installation given in detail. Four circuits have been identified for you to develop further. The worked example throughout this book is for a cooker circuit supplying one appliance within the kitchen. By following the steps used for the worked example you should be able to apply the same principles to the other circuits and eventually have a project which is fully designed. As you follow through the worked example, you should also complete the appropriate stage for Circuit 2, the reception area lighting circuit. In Chapter 3 details have been given that may apply to all of the circuits. At the end of Chapters 3, 4, 5 and 6 space has been left for you to complete the calculation for Circuit 2.

Circuits 3 and 4 should be completed after Chapter 6. The routes for these circuits should be chosen by you taking into account all the relevant factors.

It is intended to cover only the circuits detailed at the start of the project, but you may continue with your design to cover all the circuit requirements.

Questions related to the inspection and testing of the project installation are covered at the end of Chapter 8.

CT Manufacturing Ltd.

Specification

The drawings show a factory which manufactures small electrical products.

Building construction

The building is of steel frame construction with facing bricks externally and an inner skin of lightweight concrete blocks. The roof over the workshop area and half of the canteen is a double pitched roof supported by steel roof trusses having lattice girders which, in turn, are supported by stanchions positioned as shown in the drawings. The roof is of corrugated sheets with lights fitted over the workshop area. The valley between the double pitch acts as a fire escape from the first floor offices.

The height from the finished floor level to the underside of the horizontal joists is 4.0m.

The floor is a cast-in-situ concrete slab with screed finish. The first floor only covers half of the ground floor area. The roof to this is cast-in-situ concrete slabs covered with standard bitumen to the required thickness.

Electrical installation

The electrical installation is in accordance with the Electricity at Work Regulations 1989 and BS 7671:1992 Requirements for Electrical Installations.

Electricity supply

The supply is three-phase four wire 400/230 V 50 Hz.

The supply and installation form a TN-C-S system protected in the supply company's cut out by 3×100 A BSEN 60269-1:1994 (BS88 Part 2 and Part 6) type fuses.

Z_e is 0.3 Ω and I_s is 16 kA.

Wiring

The submain between the two distribution boards is a 35 mm^2 two-core XLPE/SWA/LSF cable with copper conductors which is 28 metres long, run cleated to a perforated steel cable tray and no factors apply. Circuits supplied from the first floor distribution board must not exceed 3.5 V drop. Measured from the first floor distribution board Z_{db} is 0.8242 Ω and I_s is 5.4 kA. The submain is protected by a 63 A BSEN 60269-1:1994 (BS88 Part 2 and Part 6) type fuse. Power supplies to the main workshop area are taken from an overhead busbar trunking as shown.

Lighting to the ground floor is by PVC insulated single core cables with copper conductors installed in heavy gauge B.E. steel conduit. Ground floor power is via a steel trunking and conduit system using PVC insulated single cables with copper conductors.

First floor circuits are installed in heavy gauge PVC conduit using single core PVC insulated single cables with copper conductors.

Where steel conduit or trunking is used it forms the only cpc for the circuitry.

The power to the boiler room and the whole of the fire alarm system are installed in MICC cable with copper conductors and sheath. The cable is served overall with PVC insulation and installed on steel cable tray.

Emergency lighting units are self contained, non-maintained and supplied normally from the appropriate lighting circuit to keep their batteries charged. For the calculation of circuit loads these may be ignored.

Mounting heights for electrical equipment

From F.F.L. to centre

Socket outlets		1 m
Light switches		1.4 m
Kitchen equipment		1.2 m
Distribution boards	ground floor	3 m
	first floor (ceiling height 2.4 m)	2 m

Circuit protection

With the exception of the first floor circuits HBC fuses to BSEN 60269-1:1994 are used throughout. Each of the first floor circuits is protected by a Type C MCB to BSEN 60898.

CT Manufacturing Ltd.

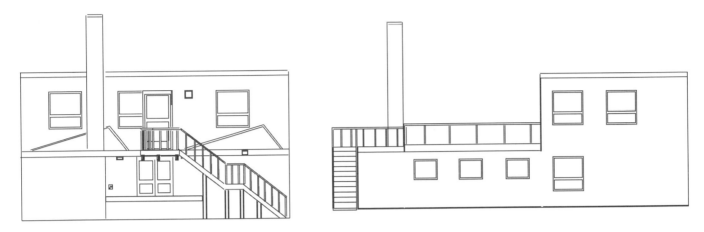

West elevation

South elevation

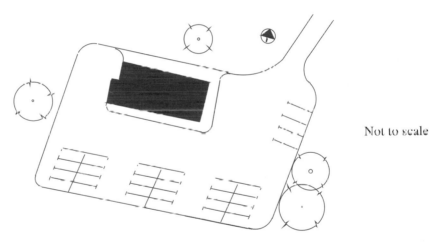

Not to scale

Site plan

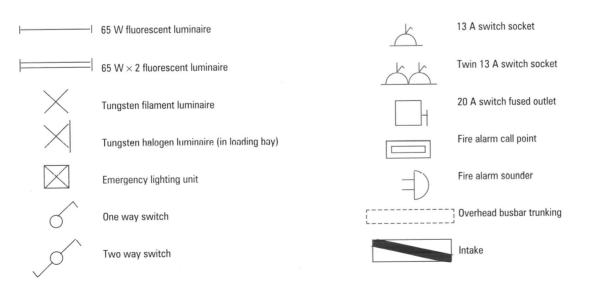

⊢————⊣	65 W fluorescent luminaire	
⊢══════⊣	65 W × 2 fluorescent luminaire	
✕	Tungsten filament luminaire	
⊗	Tungsten halogen luminaire (in loading bay)	
⊠	Emergency lighting unit	
⌀⌐	One way switch	
⌀⌐	Two way switch	

	13 A switch socket	
	Twin 13 A switch socket	
	20 A switch fused outlet	
	Fire alarm call point	
	Fire alarm sounder	
	Overhead busbar trunking	
	Intake	

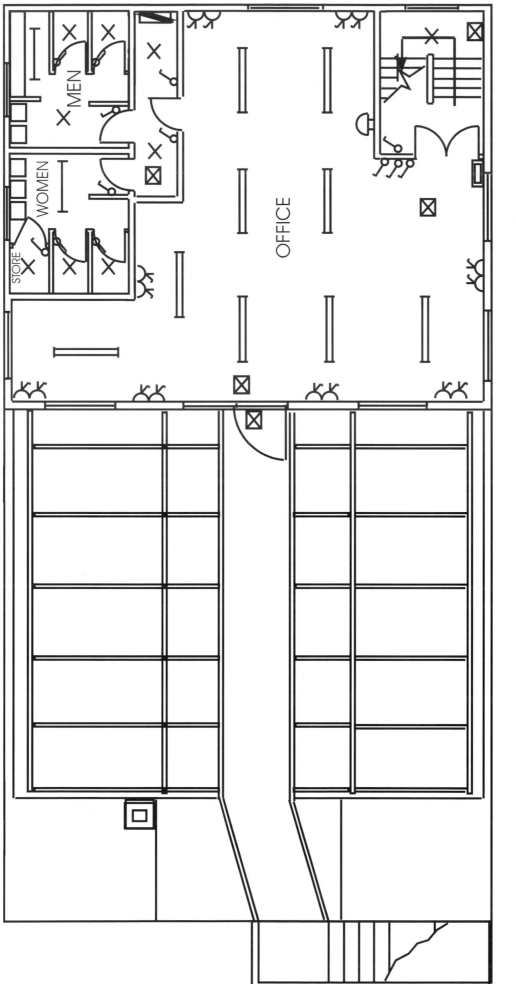

MEN

WOMEN

STORE

OFFICE

Scale 1:100

FIRST FLOOR PLAN

138

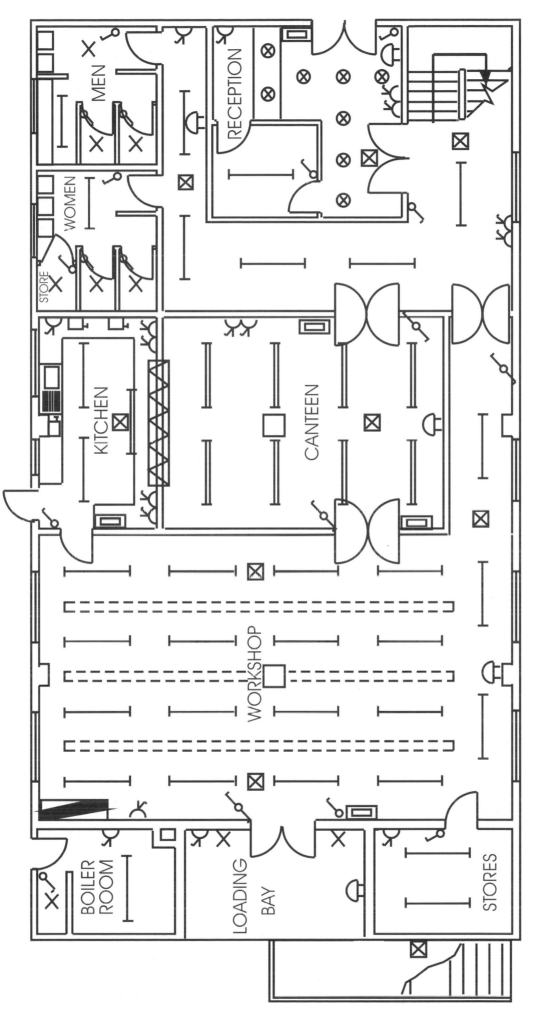

GROUND FLOOR PLAN

Scale 1:100

Circuits covered by this project are:

Circuit 1 is used as a worked example. Complete Circuit 2 as you work through the sections in this book. Circuits 3 and 4 can be completed after Chapter 6.

1. **COOKER CIRCUIT**
   Wired throughout in 32mm heavy gauge steel conduit using single core PVC cables with copper conductors and the conduit forming the circuit protective conductor. There is no socket outlet included. (Worked example)

2. **RECEPTION AREA LIGHTING CIRCUIT**
   Wired throughout in 25mm heavy gauge steel conduit using PVC cables with copper conductors and the conduit forming the circuit protective conductor. The circuit will be grouped with 3 other circuits and they do not pass through the kitchen.

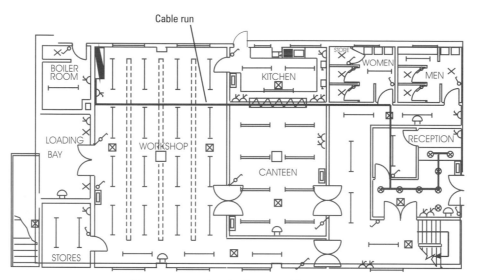

Cable run

Not to scale

Measurements to be taken from ground floor scale drawing

Cable routes for Circuits 3 and 4 are to be chosen by you and marked on the drawing.

Once this is done, the chosen route will be used to determine circuit length.

3. **FIRST FLOOR OFFICE LIGHTING CIRCUIT**
   Wired throughout in heavy gauge PVC conduit using single core PVC cable with copper conductors. Grouped with 1 other circuit.

4. **BOILER ROOM POWER**
   Wired in MICC cable with copper conductors and sheath with an overall serving of PVC. The load for this circuit is to be assumed as 15 A with the boiler room temperature at 40 °C. The cable will be clipped to a perforated steel cable tray and not grouped with other circuits.

Cable selection (Check List)

The selection of a cable for a particular circuit can be considered as a sequence of steps:

1. Determine the design current of the circuit. I_b
2. Determine the rating of the overcurrent protection device. I_n
3. Determine the factors that apply.   i.e. C_a, C_g and C_f
4. Determine the minimum current carrying capacity of the conductor. I_t
5. Determine the reference method.
6. Select the cable size from the appropriate table based on current carrying capacity. I_z
7. Ensure cable selected complies with the voltage drop constraint.
8. Determine the maximum value of earth fault loop impedance from the tables. Z_s
9. Calculate the actual value of Z_s for the circuit.
10. Ensure the circuit complies with the shock protection constraint.
11. Calculate the earth fault current. I_f
12. Establish the disconnection time from the time/current characteristics.
13. Calculate the minimum cross sectional area of the circuit protective conductor.
14. Verify that the circuit complies with the thermal constraints.

Record of Electrical Installation and Contractor

Details of the Electricity supply

Voltage . . . .

Phases . . . .

Frequency . . . .

Prospective short circuit current at origin of the installation . . . . k A

Type of earthing arrangements (TN-S, TM-C-S or TT) . . . .

Type(s) of protective devices (overcurrent protective devices) . . . .

Earth fault loop impedance at the origin . . . .

Any special installations . . . .

Details of Premises being tested

Details of Electrical Contractor

Name . . . .

Address . . . .

Name of person completing the report

. . . .

Schedule of Electrical Circuits

Distribution board reference		Location

Circuit reference	Protection device rating	Description of circuit																										

Record of Test Results

Installation:

Contractor:

Engineer:

Date of tests:

Distribution board ref:

External impedance (Z_{db}):

Minimum insulation resistance:

Supply voltage:

Instruments used	Make	Serial No.
Continuity		
Insulation		
Loop impedance		
RCD		

Circuit reference	Radial or ring	Protection device			Phase (colour or 3 phase)	Conductor size		Continuity			Insulation resistance MΩ		Polarity (pass or fail)	Earth loop impedance Ω		RCD			
		Device	Type	Rating		Live	cpc	L-L	N-N	cpc	Phase to Neutral	To earth		Design	Actual	Rated trip current (mA)	50% Y/N	Trip time (ms) at 100%	Trip time (ms) at 150 mA